T0205548

Smart Innovation, Systems and Technologies

359

Series Editors

Robert J. Howlett, *KES International Research, Shoreham-by-Sea, UK*
Lakhmi C. Jain, *KES International, Shoreham-by-Sea, UK*

The Smart Innovation, Systems and Technologies book series encompasses the topics of knowledge, intelligence, innovation and sustainability. The aim of the series is to make available a platform for the publication of books on all aspects of single and multi-disciplinary research on these themes in order to make the latest results available in a readily-accessible form. Volumes on interdisciplinary research combining two or more of these areas is particularly sought.

The series covers systems and paradigms that employ knowledge and intelligence in a broad sense. Its scope is systems having embedded knowledge and intelligence, which may be applied to the solution of world problems in industry, the environment and the community. It also focusses on the knowledge-transfer methodologies and innovation strategies employed to make this happen effectively. The combination of intelligent systems tools and a broad range of applications introduces a need for a synergy of disciplines from science, technology, business and the humanities. The series will include conference proceedings, edited collections, monographs, handbooks, reference books, and other relevant types of book in areas of science and technology where smart systems and technologies can offer innovative solutions.

High quality content is an essential feature for all book proposals accepted for the series. It is expected that editors of all accepted volumes will ensure that contributions are subjected to an appropriate level of reviewing process and adhere to KES quality principles.

Indexed by SCOPUS, EI Compendex, INSPEC, WTI Frankfurt eG, zbMATH, Japanese Science and Technology Agency (JST), SCImago, DBLP.

All books published in the series are submitted for consideration in Web of Science.

Alfred Zimmermann · R. J. Howlett ·
Lakhmi C. Jain
Editors

Human Centred Intelligent Systems

Proceedings of KES-HCIS 2023 Conference

 Springer

Editors
Alfred Zimmermann
Reutlingen University
Reutlingen, Germany

R. J. Howlett
KES International Research
Shoreham-by-Sea, UK

Lakhmi C. Jain
KES International
Selby, UK

ISSN 2190-3018   ISSN 2190-3026 (electronic)
Smart Innovation, Systems and Technologies
ISBN 978-981-99-3426-3   ISBN 978-981-99-3424-9 (eBook)
https://doi.org/10.1007/978-981-99-3424-9

© The Editor(s) (if applicable) and The Author(s), under exclusive license
to Springer Nature Singapore Pte Ltd. 2023

This work is subject to copyright. All rights are solely and exclusively licensed by the Publisher, whether
the whole or part of the material is concerned, specifically the rights of translation, reprinting, reuse of
illustrations, recitation, broadcasting, reproduction on microfilms or in any other physical way, and transmission
or information storage and retrieval, electronic adaptation, computer software, or by similar or dissimilar
methodology now known or hereafter developed.
The use of general descriptive names, registered names, trademarks, service marks, etc. in this publication
does not imply, even in the absence of a specific statement, that such names are exempt from the relevant
protective laws and regulations and therefore free for general use.
The publisher, the authors, and the editors are safe to assume that the advice and information in this book
are believed to be true and accurate at the date of publication. Neither the publisher nor the authors or the
editors give a warranty, expressed or implied, with respect to the material contained herein or for any errors
or omissions that may have been made. The publisher remains neutral with regard to jurisdictional claims in
published maps and institutional affiliations.

This Springer imprint is published by the registered company Springer Nature Singapore Pte Ltd.
The registered company address is: 152 Beach Road, #21-01/04 Gateway East, Singapore 189721, Singapore

Preface

This volume contains the proceedings of the KES-HCIS 2023, the International Conference on Human-Centred Intelligent Systems, as part of the multi-theme conference KES Smart Digital Futures 2023, held on 14–16 June 2023 in Rome, Italy, and virtually extended as a hybrid conference. We have gathered a multidisciplinary group of contributors from both research and practice to discuss the way human-centred intelligent systems are architected, modelled, constructed, verified, tested, and applied in various domains today.

Human-centred intelligent systems (HCIS) are information systems that apply artificial intelligence to support and interact with humans. Intelligent systems are now playing an important role in digital transformation in many areas of science and practice. Artificial intelligence defines core techniques in modern computing that are leading to a rapidly growing number of intelligent digital services and applications in practice. The study of HCIS includes a deep understanding of human practices and the study of human–system interaction and co-adaptation, the human-centred perspective of artificial intelligence, intelligent value creation, value-based digital models, generative AI, human-centred interaction and contexts, ethics and transparency of knowledge processes and algorithmic reasoning, along with intelligent digital architectures and engineering to support intelligent services and systems, and the digital transformation of enterprises. HCIS specifically considers the human work involved in supporting digital services and building intelligent systems, which consists of optimizing knowledge representation algorithms, collecting and interpreting data, and even deciding what and how to model data and intelligent systems.

All submissions were multiple peer-reviewed by at least two members of the International Programme Committee. We have finally accepted 16 high-quality scientific publications to be included in this proceedings volume. The major areas are organized as follows:

- Human-centred intelligent systems,
- Intelligent transportation and smart city systems,
- Edge computing technologies for mobile computing and Internet of Things, and
- Digital enterprise architecture in manufacturing, financial, and others.

We are satisfied with the quality of the programme and would like to thank the authors for choosing KES-HCIS 2023 as a forum for presentation of their work. Also, we gratefully acknowledge the hard work of the members of the International Programme Committee and the Organization team.

Alfred Zimmermann
Rainer Schmidt
Robert J. Howlett
Lakhmi C. Jain

Organization

Honorary Chairs

Toyohide Watanabe   Nagoya University, Japan
Lakhmi C. Jain   Liverpool Hope University, UK

Executive Chair

Robert J. Howlett   'Aurel Vlaicu' University of Arad, Romania and
Bournemouth University, UK

General Chairs

Alfred Zimmermann   Reutlingen University, Germany
Rainer Schmidt   Munich University of Applied Sciences, Germany

Programme Chairs

Yoshimasa Masuda   Carnegie Mellon University, USA, and Keio
University, Japan
Abdellah Chehri   Royal Military College of Canada, Canada

International Programme Committee

Imran Ahmed   Anglia Ruskin University, UK
Ahmad Taher Azar   Prince Sultan University, Saudi Arabia
Monica Bianchini   Università di Siena, Italy
Karlheinz Blank   arborsys, Germany
Gloria Bordogna   CNR-IREA, Italy
Giacomo Cabri   University of Modena and Reggio Emilia, Italy
Abdellah Chehri   Royal Military College of Canada
Dinu Dragan   University of Novi Sad, Serbia
Margarita N. Favorskaya   Reshetnev Siberian State University of Science
and Technology, Russia

Christos Grecos	Arkansas State University, USA
Vincent Hilaire	Université de technologie de Belfort Montbéliard, France
Katsuhiro Honda	Osaka Metropolitan University, Japan
Assoc. Emilio Insfran	Universitat Politècnica de València, Spain
Rashmi Jain	Montclair State University, USA
Gwanggil Jeon	Incheon National University, South Korea
Mustafa Asim Kazancigil	Yeditepe University, Turkey
Ayoub Khan	University of Bisha, Saudi Arabia
Boris Kovalerchuk	Central Washington University, USA
Chengjun Liu	New Jersey Institute of Technology, USA
Mihaela Luca	Institute of Computer Science, Romanian Academy
Yoshimasa Masuda	Carnegie Mellon University, USA/Tokyo University of Science, Keio University, Japan
Lyudmila Mihaylova	University of Sheffield, UK
Michael Möhring	Reutlingen University, Germany
Vincenzo Moscato	Università degli Studi di Napoli Federico II, Italy
Radu-Emil Precup	Politehnica University of Timisoara, Romania
Patrizia Ribino	National Research Council of Italy
Rachid Saadane	Hassania School of Public Works, Morocco
Milos Savic	University of Novi Sad, Serbia
Rainer Schmidt	Munich University, Germany
Stefano Silvestri	ICAR-CNR, Italy
Milan Simic	RMIT University, Australia
Andreas Speck	University of Kiel, Germany
Maria Spichkova	RMIT University, Australia
Eulalia Szmidt	Polish Academy of Sciences
Taketoshi Ushiama	Kyushu University, Japan
Alfred Zimmermann	Reutlingen University, Germany

Contents

Edge Computing Technologies for Mobile Computing and Internet of Things (3rd Edition)

Digital Enterprise Architecture for Human-Centric Intelligent Systems in Manufacturing, Financial, and Others

About the Editors

Alfred Zimmermann is Professor at Reutlingen University, Germany. He is Director of Research and Speaker of the Doctoral Programme for Services Computing at the Herman Hollerith Center, Boeblingen, Germany. His research is focussed on digital transformation and digital enterprise architecture with decision analytics in close relationship with digital strategy and governance, software architecture and engineering, artificial intelligence, data analytics, Internet of Things, services computing, and cloud computing. He graduated in Medical Informatics at the University of Heidelberg, Germany, and obtained his Ph.D. in Informatics from the University of Stuttgart, Germany. Besides his academic experience, he has a strong practical background as Technology Manager and Leading Consultant at Daimler AG, Germany. Professor Zimmermann keeps academic relations of his home university to the German Computer Science Society (GI), the Association for Computing Machinery (ACM), and the IEEE, where he is part of specific research groups, programmes, and initiatives. He serves in different editorial boards and programme committees and publishes results from his research at conferences, workshops, as well as in books and journals. Additionally, he supports industrial cooperation research projects and public research programmes.

R. J. Howlett is Executive Chair of KES International, a non-profit organization that facilitates knowledge transfer and the dissemination of research results in areas including intelligent systems, sustainability, and knowledge transfer. He is Visiting Professor at Bournemouth University in the UK. His technical expertise is in the use of intelligent systems to solve industrial problems. He has been successful in applying artificial intelligence, machine learning, and related technologies to sustainability and renewable energy systems; condition monitoring, diagnostic tools, and systems; and automotive electronics and engine management systems. His current research work is focussed on the use of smart microgrids to achieve reduced energy costs and lower carbon emissions in areas such as housing and protected horticulture.

Dr. Lakhmi C. Jain, PhD, ME, BE(Hons), Fellow (Engineers Australia), is with the University of Technology Sydney, Australia, and Liverpool Hope University, UK. Professor Jain serves the KES International for providing a professional community the opportunities for publications, knowledge exchange, cooperation, and teaming. Involving around 5,000 researchers drawn from universities and companies worldwide, KES facilitates international cooperation and generates synergy in teaching and research. KES regularly provides networking opportunities for professional community through one of the largest conferences of its kind in the area of KES.

Human-Centred Intelligent Systems

Leveraging Open Innovation Practices Through a Novel ICT Platform

Emmanuel Adamides⬤, Nikolaos Giarelis⬤, Nikos Kanakaris⬤,
Nikos Karacapilidis(✉)⬤, Konstantinos Konstantinopoulos, and Ilias Siachos⬤

University of Patras, 26504 Rio Patras, Greece
`karacap@upatras.gr`

Abstract. This paper reports on the development and preliminary evaluation of a novel online platform that facilitates and augments diverse open innovation practices in contemporary organizations. The proposed platform offers a friendly and sustainable solution that fully supports the processes of collection, dissemination, organization, synthesis and utilization of knowledge that comes from both the internal and external environment of an organization. It is based on prominent artificial intelligence and natural language processing technologies to meaningfully cluster and aggregate stakeholders' positions on the issues under consideration, the ultimate aim being to advance informed decision-making in the underlying data intensive and cognitively complex settings. Moreover, the platform may enable argumentation across the overall innovation development process, from idea formation to its market entry and commercialization.

Keywords: Open Innovation · software platform · decision making · argumentation · artificial intelligence · natural language processing

1 Introduction

Open Innovation has been defined as "the use of purposive inflows and outflows of knowledge to accelerate internal innovation, and expand the market for external use of innovation, respectively" [6]. By adopting open innovation practices, an organization's innovation management process becomes porous [24], in that ideas, concepts, as well as designs of products and services, flow in and out of its boundaries. In such a way, diverse knowledge sources associated with internal and external organization actors (i.e., managers, users/customers, employees, suppliers, competitors, researchers and regulators) become interconnected in many different ways, and information and knowledge items of various forms flow between them, being transformed for the development of products and services [23].

Due to the complexity of these processes in the context of pluralistic and distributed modern organizations, information and communication technologies (ICT) have a critical role to play in both engaging stakeholders and augmenting the quality of open innovation activities. Indeed, several ICT-based open innovation platforms are already available [3], while the use of ICT in open innovation implementations has been examined from different perspectives (see, for instance, [5, 7, 8, 13]). In any case, existing

© The Author(s), under exclusive license to Springer Nature Singapore Pte Ltd. 2023
A. Zimmermann et al. (Eds.): KES-HCIS 2023, SIST 359, pp. 3–12, 2023.
https://doi.org/10.1007/978-981-99-3424-9_1

approaches mainly concern the front-end of the innovation process, in that they focus on the interaction of the organization with external parties and pay little attention to knowledge (co-)creation and integration [18].

Aiming to address the above issues, while building on the synergy of human and machine reasoning [16], the platform described in this paper is based on prominent AI/NLP technologies to analyze and aggregate the content of the ideas and positions expressed by diverse types of stakeholders involved in diverse open innovation processes. The proposed solution aims to exploit and meaningfully integrate internal and external data, thus sustaining and inspiring better-informed collaboration towards innovative actions. Its key features are that (i) it facilitates the practice of argumentation-driven knowledge management in open innovation, and (ii) it builds on a knowledge graph-based representation of open innovation processes enriched with state-of-the-art discourse analysis and position summarization mechanisms to turn data intensive, unstructured user-generated content into actionable knowledge.

The remainder of this paper is organized as follows: Sect. 2 reports on related ICT solutions aiming to stress their advantages and disadvantages, and justify the rationale followed in our approach. The proposed solution is described in Sect. 3, paying particular attention to its major services, namely argumentation-based collaboration, discourse analysis and position summarization. Section 4 presents the outcomes of a preliminary evaluation. Finally, Sect. 5 outlines some concluding remarks and sketches future work directions.

2   ICT-Based Open Innovation Platforms

As mentioned in the previous section, a series of ICT-based open innovation platforms is already available. These primarily aim to facilitate the incorporation of external and internal sources in the ideation, crowdsourcing and optimization of an organization's product or service, while also providing support for tasks such as finding suitable employees or services to undertake an organization's administrative issues or specific projects. A brief analysis of state-of-the-art open innovation platforms is given below.

100% Open (https://www.100open.com/) is an open innovation platform aiming to help organizations develop new technologies and services. The platform integrates a strategy planning service, where social listening methods are incorporated to gather insights about what people desire, as well as crowdsourcing services, where companies can engage with customers to create new concepts, generate ideas and organize competitions to extract new technologies, enabling stakeholders to argue in a collaborative environment.

BrightIdea (https://www.brightidea.com/) is designed to enhance the performance of diverse open innovation processes. Its services include a workspace where organizations can engage employees and colleagues in various stages of the innovation pipeline, an "idea box" where internal and external sources can be combined for the formulation of innovative concepts, as well as a set of microsites that can host a plethora of activities, such as hackathons and naming selection competitions.

QMarkets (https://www.qmarkets.net/) provides different sub-platforms, according to the necessities and demands of each organization; namely, the *Q-ideate* platform is addressed to the development and management of innovative programs between employ-ees, *Q-scout* is a means for businesses to keep track of their products or services, *Q-open* allows the participation of partners and customers in the design and development of prod-ucts, while *Q-optimize* aims to amplify an organization's performance by leveraging the collective intelligence of the employees and utilizing analytics to measure results.

Imaginatik (https://www.imaginatik.com/) claims to provide an end-to-end platform for innovation management. The platform integrates three complementary modules for the development, the optimization and the evaluation of an idea. Through the platform's *Innovation Central* module, managers can ignite questions and competitions aiming at the collection of ideas either with an inbound or an outbound scope; the *Discovery Suite* module may help managers to further elaborate and refine an idea; finally, the *Results Engine* module may generate analytics, metrics and draw conclusions, while also facilitating project management and portfolio tracking.

Viima (https://www.viima.com/) helps organizations to gather ideas from their employees, customers and other stakeholders, and then refine and develop them in a collaborative mode. Moreover, the platform provides tools for prioritizing ideas and picking the right ones to progress, as well as tools for analyzing the innovation process to find and eliminate bottlenecks. It also demonstrates a flexible engine for evaluating ideas, enabling organizations to choose their own metrics and pick the right people to rate ideas on them. Finally, it offers a set of visualization tools to facilitate informed decision making.

The abovementioned platforms adopt a versatile approach, aiming to offer solutions for many different functions and activities related to open innovation in an organization. On the other hand, there are many platforms that focus on certain tasks. For instance, *Chaordix* (https://chaordix.com/) is an open innovation platform specialized in crowd-sourcing, assisting organizations in the process of extracting new ideas and assessing the feedback of the community; *Crowdspring* (https://www.crowdspring.com/) enables collaboration between organizations and graphic designers for a series of projects con-cerning custom designs; *Ennomotive* (https://www.ennomotive.com/) enables compa-nies publish their open innovation challenges to collaborate and obtain solutions from a big network of startups and engineers in a variety of fields.

In any case, the existing open innovation platforms lack argumentation functionali-ties, which would encourage stakeholders to contribute what they know by postulating their positions, providing them with the means to support their arguments by using an appropriate justification schema [2]; in addition, they do not cope well with data intensive settings, where numerous ideas and positions are expressed. Overall, they do not demonstrate machine reasoning features that could automatically organize, analyze and summarize the content of these ideas and positions, thus facilitating the underlying knowledge management and decision making processes [14].

3   The Proposed Solution

The development of the proposed open innovation platform follows the *Design Science* paradigm, which "seeks to extend the boundaries of human and organizational capabil-ities by creating new and innovative artifacts" [12]. This paradigm has been extensively adopted in the development of Information Systems in order to address diverse wicked problems, i.e. problems characterized by unstable requirements and constraints based on ill-defined contexts, inherent flexibility to change design processes and artifacts, and a critical dependence upon human cognitive and social abilities to produce effective solutions.

The elicitation of requirements has adopted contemporary approaches [10, 17]. It starts with considering and analyzing alternative scenarios and existing work practices, through interviews and dedicated workshops with potential users, aiming to understand how they participate in open innovation practices individually or collaboratively, as well as the relevant decision making processes. Essentially, through the above means, users were asked to describe their vision of the proposed solution. Valuable feedback obtained through them helped us shape two questionnaires designed specifically for the needs of the project; the first of them intended to investigate the requirements of potential stakeholders from an online open innovation platform (feedback from 151 stakeholders was collected), while the second one to investigate the attitude and demands of business executives from an information system that supports and facilitates open Innovation tasks (feedback from 148 middle and senior managers was collected). The analysis of the interviews, workshop discussions and questionnaires led to the definition of the requirements and specifications of the proposed solution.

In the rest of this section, we describe the major services of the proposed open innovation platform, namely argumentation-based collaboration, discourse analysis and position summarization. These services build on and extend prominent artificial intelligence and natural language processing tools and technologies to facilitate knowledge management and decision making in the setting under consideration.

3.1 Argumentation-Based Collaboration

It has been broadly admitted that argumentation-based collaboration plays an invaluable role in innovation, in that it fosters a knowledge-driven culture and augments the processes of tacit knowledge creation, sharing and leverage [9]. The service builds on broadly accepted argumentation formalisms to represent the concepts of the *problem/issue* in hand, the *proposals/positions* for its solution, and the *pro or contra arguments* related to proposals [15]. We adopt an incremental formalization approach, i.e., a stepwise and controlled evolution from a mere collection of individual ideas and resources to the production of highly contextualized and interrelated knowledge artifacts (Fig. 1).

This service augments sense-making through advanced visualization and monitoring dashboards that offer an informative and user-friendly overview of the underlying argumentation process in terms of participants' engagement and contributed knowledge, while also providing insights about the structure, evolution and dynamics of the collaboration. In addition, it enables advanced knowledge exchange and co-creation functionalities by offering a deliberation environment that supports interpretation of diverse knowledge items and their interrelationships. Moreover, it is geared towards facilitating collective decision making and consensus building through novel virtual workspaces that enable participants to assess alternative solutions.

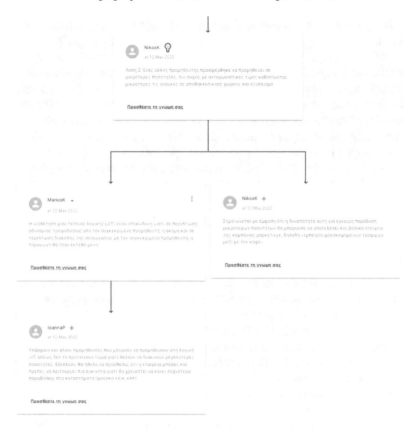

Fig. 1. A small part of an instance of argumentation-based collaboration.

3.2 Discourse Analysis

The discourse analysis service takes as input the discourse items and a set of pre-trained word embeddings. It then generates a discourse graph based on the graph-of-docs text representation [11], which facilitates three major tasks: (i) the identification of similar nodes (positions); (ii) the categorization of nodes into a set of predefined classes, using a supervised model, and (iii) the clustering of the nodes into groups, using an unsupervised model. We note here that the last two tasks produce similar outputs, i.e. groups of similar nodes; the unsupervised model is used in cases where a sufficient amount of labeled data for training is missing.

In particular, to identify similar nodes, we first calculate a list of metrics, such as *Adamic Adar* [1], *Total Neighbors* and *graph kernels*. This enables us to extract features for each node of the discourse graph, which take into consideration both the connectivity and the neighborhood characteristics of each node. By combining structure-related and text-related features of each node, we convert the node classification into a link prediction problem, which in turn can be easily converted into a binary classification problem.

As far as the task of categorizing similar nodes into a set of predefined classes is concerned, we initially generate a word and a document similarity graph. Each of them is used to perform representation learning, aiming to enrich the already existing word and document embeddings. To do so, we employ widely adopted techniques for generating node embeddings, such as *DeepWalk, Node2Vec* and *FastRP* [20]. Moreover, to reduce the feature space of our dataset, we perform a feature selection process, which identifies communities of similar documents and selects the *top-N* most representative words of each community; these words are used as features from the classification model.

Finally, for the node clustering task, we first split the nodes into groups based on the language of their text, i.e. those containing Greek and those containing English text. We further split the groups of documents into subgroups by using the *k-means* algorithm, which takes as input the document embeddings. To determine the number of groups k, we utilize the *elbow* method [25].

3.3 Position Summarization

As argued above, one of the basic capabilities of an open innovation platform should be to automatically summarize related positions and appropriately reveal their most prominent concepts. To develop our position summarization service, we develop a multistep Natural Language Processing (NLP) pipeline, which utilizes the pre-trained English large language model of *spaCy* (https://spacy.io/). The first step is to remove stop words and assign part of speech tags to each position inserted in the pipeline, in order to keep only the tokens that are nouns or adjectives. The second step is to retrieve the pre-trained word embedding representation for each token from the aforementioned model, which is used to calculate the mean embedding vector, namely the document embedding for each position. The third step is to compare the document embeddings for each position in a pairwise manner using cosine similarity. We assume that if the cosine similarity is more than a cut-off value of 50%, the associated pair of positions can be considered as similar. As a fourth step, we construct a knowledge graph, where each node represents a position and the edges between positions connect similar ones (we also store the similarity cosine percentage, as a weight of the edge). This graph is then utilized from a community detection algorithm, namely *Louvain* [4], which is used to find graph communities of thematically similar positions.

As a next step, for each graph community found, we extract its summary from the combined text, by applying the *TextRank* approach [19], which is built in the *pyTextRank* library (https://derwen.ai/docs/ptr/). This library develops TextRank as a spaCy component, which easily allows us to append it in our spaCy NLP pipeline. The resulting summaries can be organized by position type (pro or contra) and their thematic aspect. If appropriate, an overall summary from all graph communities can be also extracted (Fig. 2). It is finally noted that the TextRank approach additionally allows us to extract the *top-n* keyphrases from the entire discussion, thus quickly identifying its topics.

Fig. 2. An instance of the position summarization service (the top part corresponds to the summary of the related positions; the extracted keyphrases are listed in the bottom part).

4 Preliminary Evaluation

The proposed open innovation platform has been already evaluated by two different groups of users, one comprising employees from two large organizations operating in the food and beverages sector in Greece, and another formed by well-experienced researchers in the area of information systems who are very familiar with contemporary business applications (these groups consist of 16 and 9 users, respectively). A use case of the proposed platform, which covers all three services discussed in the previous section, was presented to the evaluators by an evaluation manager who also answered questions and provided clarifications about diverse functionalities of the platform.

The aim of this preliminary evaluation was to assess the accessibility, acceptance, overall quality, usability and ease-of-use of each service. As far as the accessibility, acceptability and overall quality of the services are concerned, the evaluators were asked to rank a series of declarations (shown in the left part of Fig. 3) using a Likert scale from 1 to 5, where 1 corresponds to "strongly disagree" and 5 to "strongly agree". Evaluators were also encouraged to leave comments on each declaration, aiming to get insights towards improving the proposed services. Figure 3 illustrates the medians calculated for each declaration and service, which reveal the aspects of services that work well or need amendments. As it can be easily noticed, the evaluators admitted that all services were novel and helpful to the work they had to perform in the context of the particular open innovation use case. However, it was made obvious that the interfaces of all services need to be improved.

Evaluators were also asked to assess the ease-of-use and the usability of the three services. In particular, they were asked to rank a series of features, such as the quality of feedback and notifications provided to the user, the handling of error prevention, the ease of identifying the required actions, the handling of information overload, the quality of help and documentation provided etc. This questionnaire (shown in the left part of Fig. 4) was formed by using the concepts and framework developed in [21] and [22]. It is noted that the answers expected in this questionnaire were in the range from 1 to 10 (with 1 corresponding to "none" and 10 to "perfect"). Figure 4 summarizes the feedback (medians) for each service and feature under consideration. As it can be seen, for the majority of features, the discourse analysis service is the one that calls for amendments.

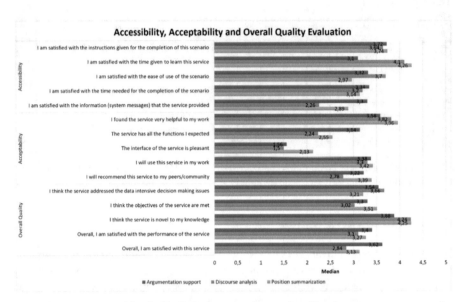

Fig. 3. Preliminary evaluation results - Part A.

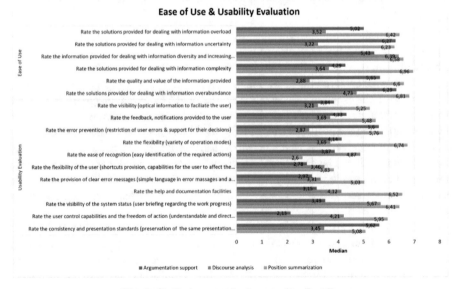

Fig. 4. Preliminary evaluation results - Part B.

5 Conclusions

This paper has described a novel software platform that facilitates and enhances diverse open innovation practices in contemporary organizations. The proposed solution builds on and extends a series of state-of-the-art artificial intelligence and natural language processing tools and technologies to meaningfully cluster and aggregate stakeholders' positions. In addition, it adopts an argumentation-based collaboration approach to augment knowledge reification and co-creation. Through dedicated services, it aims to advance knowledge management and informed decision-making in the data-intensive and cognitively-complex settings of open innovation processes. The proposed solution has been shaped through long collaboration among diverse types of stakeholders, through which a series of rich application scenarios have been designed and analyzed. Preliminary evaluation results were positive, and have justified the rationale of our approach towards integrating machine reasoning features that can automatically organize, analyze and summarize the content of ideas and positions expressed in the settings under consideration. Future work directions include the development and integration of an enhanced version of the services described in this paper, by taking into account the evaluators' initial feedback, as well as the assessment of the integrated platform in diverse open innovation settings, covering different sectors and innovation development processes.

Acknowledgements. The work presented in this paper is supported by the inPOINT project (https://inpoint-project.eu/), which is co-financed by the European Union and Greek national funds through the Operational Program Competitiveness, Entrepreneurship and Innovation, under the call RESEARCH–CREATE–INNOVATE (Project id: T2EDK-04389).

References

1. Adamic, L.A., Adar, E.: Friends and neighbors on the Web. Soc. Networks **25**(3), 211–230 (2003)
2. Adamides, E.D., Karacapilidis, N., Konstantinopoulos, K.: Argumentation schemes in technology-mediated open innovation product-service models: an activity systems perspective. Systems **9**(4), 91 (2021). https://doi.org/10.3390/systems9040091
3. Bieler, D.: The Forrester Wave: Innovation Management Solutions, Q2. Forrester (2016)
4. Blondel, V.D., Guillaume, J.L., Lambiotte, R., Lefebvre, E.: Fast unfolding of communities in large networks. J. Stat. Mech. Theory Exp. (2008). https://doi.org/10.1088/1742-5468/2008/10/P10008
5. Bugshan, H.: Open innovation using Web 2.0 technologies. J. Enterp. Inf. Manage. **28**(4), 595–607 (2015)
6. Chesbrough, H.: Open innovation: a new paradigm for understanding industrial innovation. In: Chesbrough, H., Vanhaverbeke, W., West, J. (eds.) Open Innovation: Researching a New Paradigm, pp. 1–12. Oxford University Press, Oxford (2006)
7. Corvello, V., Gitto, D., Carlsson, S., Migliarese, P.: Using information technology to manage diverse knowledge sources in open innovation processes. In: Eriksson, L.J., Wiberg, M., Hrastinski, S., Edenius, M., Ågerfalk, P. (eds.) Managing Open Innovation Technologies, pp. 179–197. Springer, Berlin (2013). https://doi.org/10.1007/978-3-642-31650-0_12
8. Cui, T., Ye, H., Teo, H.H., Li, J.: Information technology and open innovation: a strategic alignment perspective. Inf. Manage. **52**, 348–358 (2015)

9. du Plessis, M.: The role of knowledge management in innovation. J. Knowl. Manag. **11**(4), 20–29 (2007)
10. Franch X.: Software requirements patterns - a state of the art and the practice. In: Proceedings of the 37th IEEE/ACM International Conference on Software Engineering, Florence, pp. 943–944 (2015)
11. Giarelis, N., Kanakaris, N., Karacapilidis, N.: On a novel representation of multiple textual documents in a single graph. In: Czarnowski, I., Howlett, R.J., Jain, L.C. (eds.) IDT 2020. SIST, vol. 193, pp. 105–115. Springer, Singapore (2020). https://doi.org/10.1007/978-981-15-5925-9_9
12. Hevner, A.R., March, S.T., Park, J., Ram, S.: Design science in information systems research. MIS Q. **28**(1), 75–105 (2004)
13. Hrastinski, S., Kviselius, N.Z., Ozan, H., Edenius, M.: A review of technologies for open innovation: characteristics and future trends. In: Proceedings of the 43rd Hawaii International Conference on System Sciences, pp. 1–10 (2010)
14. Karacapilidis, N., Christodoulou, S., Tzagarakis, M., Tsiliki, G., Pappis, C.: Strengthening collaborative data analysis and decision making in web communities. In: Proceedings of the 23rd International Conference on World Wide Web, ACM Press, pp. 1005–1010 (2014)
15. Karacapilidis, N., Papadias, D.: A computational approach for argumentative discourse in multi-agent decision making environments. AI Commun. **11**(1), 21–33 (1998)
16. Karacapilidis, N., Rüping, S., Tzagarakis, M., Poigné, A., Christodoulou, S.: Building on the synergy of machine and human reasoning to tackle data-intensive collaboration and decision making. In: Watada, J. et al. (eds.) Intelligent Decision Technologies. Smart Innovation, Systems and Technologies, vol. 10. Springer, Heidelberg (2011) https://doi.org/10.1007/978-3-642-22194-1_12
17. Lau, L., Yang-Turner, F., Karacapilidis, N.: Requirements for big data analytics supporting decision making: a sensemaking perspective. In: Karacapilidis, N. (ed.) Mastering Data-Intensive Collaboration and Decision Making. SBD, vol. 5, pp. 49–70. Springer, Cham (2014). https://doi.org/10.1007/978-3-319-02612-1_3
18. Malhotra, A., Majchrzak, A.: Managing crowds in innovation challenges. Calif. Manage. Rev. **56**(4), 103–123 (2016)
19. Mihalcea, R., Tarau, P.: TextRank: bringing order into texts. In: Proceedings of the 2004 Conference on Empirical Methods in Natural Language Processing, pp. 404–411. Association for Computational Linguistics, Barcelona, Spain (2004)
20. Mikolov, T., Chen, K., Corrado, G., Dean, J.: Efficient estimation of word representations in vector space. arXiv preprint arXiv:1301.3781 (2013)
21. Nielsen, J.: Designing Web Usability: The Practice of Simplicity. New Riders Publishing, Indianapolis (1999)
22. Norman, D.A.: The Design of Everyday Things, Revised and Expanded Edition The MIT Press, Cambridge (2014)
23. Teece, D.J.: Strategies for managing knowledge assets: the role of firm structure and industrial context. Long Range Plan. **33**, 35–54 (2000)
24. Tidd, J., Bessant, J.: Strategic Innovation Management. Wiley & Sons, Chichester (2014)
25. Trupti, M.K., Prashant, R.M.: Review on determining number of cluster in K-Means clustering. Int. J. Adv. Res. Comput. Sci. Manage. Stud. **1**(6), 90–95 (2013)

Advanced Analytics for Smart Farming in a Big Data Architecture Secured by Blockchain and pBFT

El Mehdi Quafiq[1], Abdellah Chehri[2(✉)], and Rachid Saadane[1]

[1] SIRC-(LaGeS), Hassania School of Public Works, Casablanca, Morocco
saadane@ehtp.ac.ma
[2] Department of Mathematics and Computer Science, Royal Military College of Canada, Kingston, ON K7K 7B4, Canada
chehri@rmc.ca

Abstract. In this paper, we adopted the strategy of utilizing Blockchain technology for data processing. This concept is implemented as a unique data structure tasked with keeping historical data on creative farm transactions. Using this security mechanism, users will be able to store sensitive data and share it with one another without the need for a centralized authority. From the standpoint of federated business intelligence analytics, the Blockchain is the technological engine of crypto city data or crypto farming data. Given that Big Data is the only source of all the useful information for the critical key performance indicators of intelligent farming data analytics, Big Data is the only source of truth. In this study, we also demonstrated how the Big Data Lake and Blockchain architectures may be integrated using the design of distributed systems. In addition, we determined how this design would be implemented in the real world and detailed the functional and technological components.

Keywords: Smart Agriculture · Smart City · Data Analytics · Big Data · Blockchain · Practical Byzantine Fault Tolerance pBFT

1 Introduction

Today, climate change is a fact with immediate implications including an increase in the frequency of extreme climatic events such as droughts and floods. Due to the harmful effects of global warming on agriculture and the environment. In a climate that is continually changing, agriculture should continue to feed a growing population without damaging the environment. The intricacy of the climate change phenomenon necessitates the application of effective and diversified solutions to this issue.

Smart Farming, or more simply Agriculture 4.0, refers to the use of technology in agriculture, horticulture and aquaculture, in particular with the help of software, automation and data analysis, to improve the yield, efficiency and profitability of the agricultural sector.

© The Author(s), under exclusive license to Springer Nature Singapore Pte Ltd. 2023
A. Zimmermann et al. (Eds.): KES-HCIS 2023, SIST 359, pp. 13–23, 2023.
https://doi.org/10.1007/978-981-99-3424-9_2

In our previous works [1–6], we proposed a Big Data Architecture and a Data Migration Strategy for Smart Cities and Smart Farms, which involve sophisticated architectural components that must be executed following the smart farming technical and functional requirements.

1. *The many data sources of smart agriculture*, where IoT devices, sensors, and drones play a significant role;
2. *Data Processing in real-time and batch mode*, taking into account the variety of data sources, data volume, and the unpredictability of data speed;
3. *Data Storage* considers each data source's uniqueness in terms of data structure and type;
4. *Data Modeling* from a business intelligence perspective based on data analytics needs and data source type;
5. *ETL*, which stands for extract, transform, and load, is utilized to integrate data for long-term storage in data warehouses, data hubs, and data lakes. Typically, it is applied to known, pre-planned sources to organize and prepare data for conventional business intelligence and reporting.
6. *Data Quality in real-time* and following the execution of ETL operations;
7. *Data Science* models are created by determining at which layer of the data lake the machine learning model should be executed.

The technical limitations of data processing in a Big Data environment and Hadoop in particular were enumerated and analyzed in order to address them in the SSOBDA data architecture (Smart Systems Oriented Big Data Architecture).

The technical limitations of data processing in a Big Data environment and Hadoop in particular were enumerated and analyzed in order to address them in the SSOBDA data architecture (Smart Systems Oriented Big Data Architecture). Our primary objective was and remains to facilitate agricultural processes by providing farmers with data-driven solutions that assist them in maintaining their sustainability, spatial distribution-related processes, water management, and other processes.

The number of use cases for blockchain technology is steadily expanding in tandem with the progression of this disruptive innovation's underlying blockchain architecture. It is possible for the consensus algorithm to obtain distributed consensus among the network's nodes. At the moment, for the practical byzantine fault tolerance algorithm, also known as PBFT, which is the consensus algorithm most frequently employed in the alliance blockchain, it is necessary for all nodes in the network to take part in the process of reaching a consensus.

The primary objective of this research is to ensure: 1. A secure transmission of smart farms dedicated to advanced data analytics; 2. A remedy for the scalability flaws of blockchain by devoting data storage primarily to the Big Data Lake. Reduce burdens on the Big Data platform by processing secured data (in the form of streams and micro-batches) over the blockchain and storing the insightful data generated by the Analytics Zone of the data lake on the blockchain.

In this article, we will discuss the architecture, programming languages, APIs, and technologies that enable the deployment of an advanced analytics solution for Smart Farming in a Big Data Architecture and secure it with Blockchain and PBFT.

The following describes the format of this paper: The second section, "Sect. 2", includes an Introduction that provides a concise summary of our earlier work and explains how it led us to suggest this solution that incorporates elements of both big data and blockchain. In Sect. 3, a description is given of how advanced data analytics systems for smart cities and smart farming should be built. In Sect. 4, we discuss the architecture that has been suggested. Demonstrate in Sect. 5 how to include a big data architecture into a blockchain that is secured by PBFT. The conclusion and the article's perspective could be addressed in Sect. 4.

2   Blockchain from Data Processing Perspective

The blockchain is a storage technology with the particularity of containing a copy of each piece of information recorded in each of the nodes that compose it. This technology consists of a distributed ledger composed of a "blockchain" containing a list of activities attested and encrypted. It is a network that is shared with all the actors involved and which is decentralized, which makes it a disruptive innovation technology for a smart agriculture that, until now, stores its data on a server centralize. In addition, the blockchain is secured thanks to a system of cryptography of data, to make them immutable and - practically – falsifiable. A blockchain is a decentralized and secured database that can be built across a network to store the history of all exchanges between its users since the creation. One of the particularities of the blockchain is that it is shared by its various users, without intermediaries, which allows everyone to check the validity of the chain [16, 17].

Computer security is the field whose objective is to protect information systems from malicious. Protections are implemented at various levels, including technical, institutional, and human. As described by the ISO 27001 standard, one of the approaches to evaluate security is to establish specific properties that the information system must respect:

- Confidentiality: only authorized individuals have access to the data.

  This may be readable by an unauthorized individual, but it must remain incoherent.

- Integrity: the data have not been manipulated and are accurate and exhaustive.

  An adversary must be unable to alter or forge them.

- Authenticity: the issuer is who he states he is.
- Accessibility: the data is available when a user requires it.
- Non-repudiation: assurance that a message was transmitted by its sender and/or received by its recipient.

2.1   Blockchain Data Structure and Transactions

Even while the individual nodes in the blockchain system don't put their complete faith in one another, the network as a whole can reliably update and manage the set of shared global states and carry out the transactions that result in those changes. The blockchain is a specific data structure that stores these previous states and trades.

For a transaction to be validated, all nodes must consent to it. In the data structure of a blockchain, each block has a cryptographically generated pointer that connects it to the block that came before it, all the way back to the genesis block (also known as the first block). The blockchain is most commonly known as the distributed ledger [7–9].

The transactions in a blockchain can be viewed in the same way as traditional database transactions, which are just a sequence of operations carried out in various states. This can be observed from the perspective of database management—the rationale behind each transaction in a blockchain needs to adhere to the semantics of ACID. The failure model that must be considered stands out as the primary distinction between the two ideas.

To ensure that ACID compliance is met, the currently available transaction distributed databases use tried-and-true concurrency control methods, such as the two-phase commit. They can get a decent performance because this straightforward approach makes the simplest crash failure model possible. On the other hand, the architecture of the "native" blockchain considers a more adversarial environment in which nodes exhibit Byzantine behavior. As a result, the overhead associated with the control of concurrency is significantly increased in models of this kind [10-12].

2.2 Private and Public Blockchain

In private blockchains, access and usage are restricted to a limited number of participants. This enables internal experimentation (for a Smart Farm, for example), but with limitations on innovation (limited ecosystem) and the cost of implementing infrastructure. In contrast, the public blockchain must operate with a programmable token (or programmable currency like Bitcoin).

The network users' transactions will be grouped into blocks, and each block will be validated by network nodes known as "miners" using techniques that depend on the type of blockchain; for instance, in the Bitcoin blockchain, this technique is known as "Proof of Work" and consists of solving algorithmic problems. The block is timestamped and posted to the blockchain once it has been confirmed. The transaction is then visible to both the recipient and the network as a whole [13–16].

2.3 Practical Byzantine Fault Tolerance

The Practical Byzantine Fault Tolerance, often known as pBFT, is a communication bond that ensures both liveness and safety in networks that are only partially synchronous. It is essentially a consensus method or protocol that defends against assaults that put the network's security at risk [17–19]. It was developed to function effectively in asynchronous systems where high levels of consistency are necessary. The PBFT is a collection of protocols that are broken down into the following three stages [20]:

1. *The pre-prepare phase:* where the leader will broadcast the value that needs to be committed by the other nodes;
2. *The prepare phase:* where the nodes will broadcast the values that they will commit;
3. *The commit phase:* where the committed value will be confirmed based on the agreement of the above of the two third of the nodes in the previous phase.

3 Smart Farming and Smart City Data Analytics

In this paper, we considered the idea of "smart farming" as an essential component of the administration of "smart cities." This is because farms are a component of cities, and the goal of smart farming and smart city management is to improve the living conditions of its residents (including farmers).

The first thing that we did was identify the problems that cities and agriculture face and the data analytics solutions that can fix those problems. Then, after defining the technological hurdles posed by these challenges in terms of data storage and processing, we devised a Data Migration Strategy and designed a Big Data Architecture to accommodate the analytics requirements of smart cities and smart farms.

3.1 Business Analysis for Smart Farming

In today's world, the city's administration and agricultural practices face various issues. These difficulties may be associated with various domains and sectors, including the financial sector, mining, criminal activity, food production and health, sustainability, pandemic robustness, water management, traffic, irrigation, digitalization of administrative operations, and so on.

Because each problem calls for a unique technical solution, it is challenging to develop a data lake that satisfies the needs of both smart cities and smart farming in their whole or the majority of cases. In fact, conducting business analysis in each field or industry will take a lot of work. Therefore, we offer a method that classifies any problem or use case of the Smart Farm (that we believe to be a challenge or an agricultural requirement) into one of the following three categories:

1. Analysis based on spatial distribution that focuses mainly on climate, productivity, and global challenges (such as drought and sustainability);
2. Water management that focuses primarily on the long-term viability of water supplies and efficient irrigation;
3. Predictive maintenance of mining equipment and other types of mechanical systems, as well as their general upkeep and maintenance.

It is possible to build a unified solution for all of these different difficulties by first classifying them into distinct categories and then analyzing the business logic behind them and the qualities they share. This technique has the potential to be expanded and applied to a wider variety of smart systems and ideas beyond just smart farming.

4 Smart Systems Oriented Big Data Architecture (SSOBDA)

Big Data architecture should be built in layers to meet the data analytics needs of Smart Cities and Smart Farming. This will ensure greater safety when it comes to granting privileges to users and adding permissions to users or groups of users in different directories and schemas of the data lake.

Additionally, this will allow for greater flexibility when it comes to managing various data types and structures within the various schemas of the data lake, such as by using the Avro format, which is raw-oriented for the associated files with a given (measures and dimensions). As can be seen in Fig. 4, our proposed architecture is made up of the following elements:

1. **Share Area:** Is built on top of NFS (Network File System) as a group of folders designed based on a directory structure that meets the data analytics requirements, and it is the Landing Area for the ephemeral data (mainly DHOs) and the DGM data that is coming from less secured systems. 2. Backup Area: is built on top of NFS (Network File System) as a group of folders designed based on a directory structure that meets the data analytics requirements. In addition to this, the Shared Area is thought of as a doorway to the Hadoop platform;

2. **Raw Zone:** The location where we store raw data that has been received straight from the source in a format and structure that is analogous to the source;

3. **Structured Layer:** This is where the data will be purified and stored in partitioned tables, as opposed to the raw zone's external tables. Additionally, according to the Hybrid data model, the Historical data will be stored in this particular zone;

4. **The Trusted Zone:** Is the location where the data, in the form of fact tables and dimensions, will be stored in accordance with the logical data model. This zone will be regarded as the veritable source of information for the data analytics pertaining to smart cities and smart farming;

5. **The Enrichment Layer:** Is the point at which the data will be changed and enriched with the results of KPI calculations. Data will be stored in this layer in the form of Data Marts, Datasets, and Views, and it will be prepared to be processed by Data Visualization tools and Machine Learning algorithms, as well as by Smart Machines. These Smart Machines will carry out actions based on the outputs generated in this Enrichment Layer.

5   Architectural Requirements to Combine Big Data Architecture with Blockchain

5.1   Theoretical Physical Architecture

Figure 1 depict our proposed layout for the architecture of the system. On-premises, cloud, or hybrid infrastructure, may be utilized to construct this proposed architecture (e.g., Data Nodes, Node Manager). The only prerequisite is that the cluster should be spread out among numerous nodes, and the architecture is centered on the Primary node (which can also be known as the Name Node or the Resource Manager).

With the assistance of this distributed architecture, we can build decentralized databases (Blockchain) and distributed file systems (e.g., HDFS). Then, on top of the blockchain, we will be able to construct the PBFT-based system that will guarantee the safety of the data.

On top of Hadoop, we can install Big Data frameworks (such as MapReduce and Spark) and warehouses (e.g., Hive Warehouse). After that, if we install the API and libraries (such as Spark MLlib) that are required to build ETL and machine learning models, we will be able to do so.

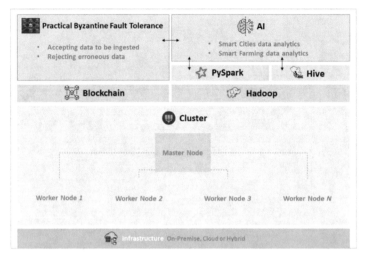

Fig. 1. Theoretical Physical Architecture

5.2 Deployed Physical Architecture

In order to deploy our theoretical solution and imitate the same behaviour of an architecture that has been implemented in a production environment, we had to set up the following components, as indicated in Fig. 2. This was necessary for us to accomplish both of these goals.

- **Prerequisites:** In order to meet the requirements, we had to install GNU/Linux Ubuntu. On top of that, we will install Python along with all of the APIs and libraries, and then we will write the scripts. The next step is to construct the nodes and identify their hosts as well as their logical and physical ports.
- **Infrastructure:** Each of the nodes is constructed on top of a containerized Docker platform that is hosted locally.
- **Big Data Architecture:** The Big Data Architecture is built on Cloudera Hadoop distribution, which is where we keep our big data components like Hive and Spark. This architecture was designed.
- **Blockchain:** Utilizing Flask to construct a web application that enables us to submit new data records by making use of urllib3 requests, together with two Python scripts:
- *manage-Chain.py*, which will allow us to process historical data from the blockchain and the big data lake, then predict the Smart City and Smart farm index, and launch the hashing mechanism in order to generate new hashes;
- *blockchain.py*, which will take care of validating the data based on the PBFT system and either reject it or ingest it in the blockchain. Manage-Chain.py will allow us to process historical data from the blockchain and the big data lake.

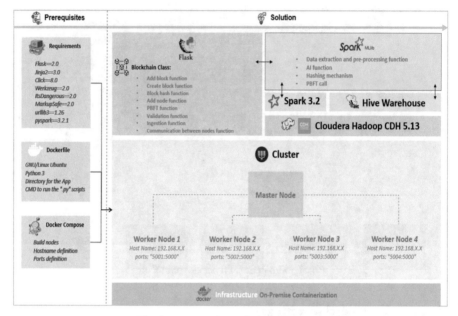

Fig. 2. Deployed Physical Architecture

5.3 Functional Architecture

As shown in Fig. 3, from a functional point of view, in order to fulfill the data analytics requirements for smart cities and smart farming, we needed to conceive of the solution paradigmatically, as well as from a variety of various viewpoints and perspectives. Therefore, we were forced to create of:

- **Data Source Layer:** This layer encapsulates a lengthy process in which data will flow from the source to either another Business Warehouse, then to the Data Cube to obtain only the significant characteristics, or it will be pushed directly to the Blockchain/Data Lake. Within this layer, we build the connectivity patterns that are based on Rest APIs, ODBC/JDBC Drivers, Connectors, and other such things. In addition, we are hard at work defining the data extraction tools (such as Sqoop, Flume, Kafka, NiFi, and Web requests) that will allow us to ingest data from sensors, web services, IoT devices, RDBMS, and other sources. These tools must be able to meet the requirements of big data in terms of batch and real-time data processing.

- **Data Engineering and AI Analytics Layer:** This is the layer where data pretreatment jobs will be conducted to import data, cleanse it, and transform it into a format ready to be analysed by business intelligence tools and machine learning algorithms. Within this investigation's scope, we used the framework Spark to partition the iterations and calculations across a number of different nodes. Therefore, the PySpark script will process historical data from the Blockchain after the preprocessing phase to train, test, and build the model. Next, it will predict the Smart City Index for the N + 1 records (that are newly imported), push them into the hashing mechanism to allocate a unique hash to each record, and it will finish by predicting the Smart City Index for the newly imported records. The data will be sent to the PBFT program to be validated after being cleaned with the corresponding hash.

- **Practical Byzantine Fault Tolerance Layer:** This is where the data will be analyzed, and the results will be sent back to the client in the form of three messages, as illustrated in Fig. 5.

1. Notification of failure, which indicates that the data are incorrect; for example, anomalies or outliers;
2. The data will not be added since it already exists in the blockchain;
3. A new node will be added when the data is valid and does not already exist in the Blockchain.

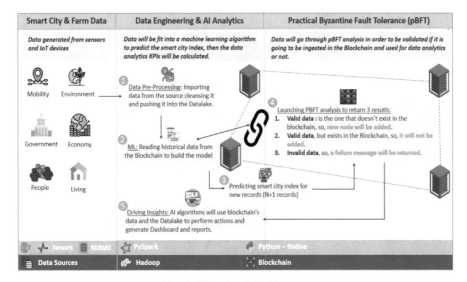

Fig. 3. Functional Architecture

As depicted in Fig. 4, data will flow from the source to the data lake via the following channels:

1. Other business data marts for cities management, government, economy, and more, then it will either land on the shared area if the source system is considered not secure or on data cube to extract only the needed columns and perform aggregations to calculate some of the main KPIs like the smart city and smart farm score in a specific area or domain like government, environment, and others before pushing it into the data lake;
2. Directly pushed to the Data Cube or Shared Area;
3. Distilled from the Data Lake;

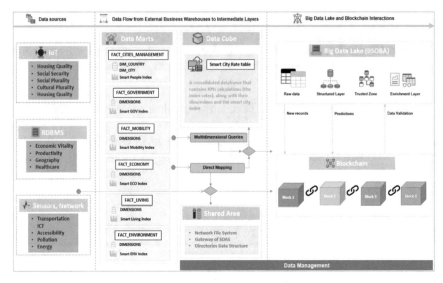

Fig. 4. Technical Architecture

6 Conclusion and Perspective

In this study, we determined the theoretical side of how we may benefit from the Master and Worker design of distributed systems in order to converge two types of architectures, namely Big Data and Blockchain, which will provide more security and less labour for worker nodes to handle. Specifically, we did this so that we could determine how we may benefit from the Master and Worker design of distributed systems.

In addition to this, we worked on the delivery of a solution that made use of Docker containers, Flask, and the Cloudera Hadoop platform. This approach covered all theoretical aspects while simulating the behavior of a production deployment.

In our future research, we plan to build an abstraction layer on top of the Big Data Lake. This layer will automatically overcome the limitations of big data environments (such as handling small files, archiving, and more) by referring to configuration tables. That will be enriched by the abstraction framework and machine learning algorithms that will learn from the system errors using the logs as historical data. In other words, the abstraction layer will automatically overcome the limitations of big data environments.

References

1. Ouafiq, E., Elrharras, A., Mehdary, A., Chehri, A., Saadane, R., Wahbi, M.: IoT in smart farming analytics, big data based architecture. In: Zimmermann, A., Howlett, R.J., Jain, L.C. (eds.) Human Centred Intelligent Systems. SIST, vol. 189, pp. 269–279. Springer, Singapore (2021). https://doi.org/10.1007/978-981-15-5784-2_22
2. Ouafiq, E.M., Saadane, R., Chehri, A.: Data management and integration of low power consumption embedded devices IoT for transforming smart agriculture into actionable knowledge. Agriculture **12**, 329 (2022). https://doi.org/10.3390/agriculture12030329

3. Ouafiq, E.M., Saadane, R., Chehri, A., Jeon, S.: AI-based modeling and data-driven evaluation for smart farming-oriented big data architecture using IoT with energy harvesting capabilities. Sustain. Energy Technol. Assessments **52**(Part A), 102093, ISSN 2213–1388 (2022). https://doi.org/10.1016/j.seta.2022.102093

4. Ouafiq, E.M., Raif, M., Chehri, A., Saadane, R.: Data architecture and big data analytics in smart cities. Procedia Comput. Sci. **207**, 4123–4131 (2022), ISSN 1877–0509. https://doi.org/10.1016/j.procs.2022.09.475

5. Ouafiq, E.M., Saadane, R., Chehri, A., Wahbi, M.: 6G enabled smart environments and sustainable cities: an intelligent big data architecture. In: 2022 IEEE 95th Vehicular Technology Conference: (VTC2022-Spring), Helsinki, Finland, pp. 1–5 (2022). https://doi.org/10.1109/VTC2022-Spring54318.2022.9860772

6. Ouafiq, E.M., Saadane, R., Chehri, A., Wahbi, M.: Data lake conception for smart farming: a data migration strategy for big data analytics. In: Zimmermann, A., Howlett, R.J., Jain, L.C. (eds.) Human Centred Intelligent Systems: Proceedings of KES-HCIS 2022 Conference, pp. 191–201. Springer, Singapore (2022). https://doi.org/10.1007/978-981-19-3455-1_15

7. Dinh, T., et al.: Untangling blockchain: a data processing view of blockchain systems. IEEE Trans. Knowl. Data Eng. **30**(07), 1366–1385 (2018)

8. Herschel, R.T. (ed.): Principles and Applications of Business Intelligence Research. IGI Global (2013). https://doi.org/10.4018/978-1-4666-2650-8

9. Information Resources Management Association, Business Intelligence: Concepts, Methodologies, Tools, and Applications (4 Volumes) (2016). https://doi.org/10.4018/978-1-4666-9562-7

10. Stonebraker, M., Madden, S., Abadi, D.J., Harizopoulos, S., Hachem, N., Helland, P.: The end of an architectural era (it's time for a complete rewrite). In: Proceedings 33rd International Conference Very Large Data Bases, pp. 1150–1160 (2007)

11. Corbett, J.C., et al.: Spanner: Google's globally-distributed database. In: Proceedings 10th USENIX Symposium Operating System Design Implementation, pp. 261–264 (2012)

12. Castro, M., Liskov, B.: Practical byzantine fault tolerance. In; Proceedings 3rd USENI

13. Junqueira, F.P., Reed, B.C., Serafini, M.: Zab: highperformance broadcast for primary-backup systems. In: Proceedings IEEE/IFIP International Conference Dependable System Network, pp. 245–256 (2011)

14. Ongaro, D., Ousterhout, J.K.: In search of an understandable consensus algorithm. In: Proceedings USENIX Annual Technical Conference, pp. 305–319 (2014)

15. Lamport, L.: Paxos made simple, fast, and byzantine. In: Proceedings of 6th International Conference Principles Distribution System, pp. 7–9 (2002)

16. Nakamoto, S.: Bitcoin: a peer-to-peer electronic cash system (2008). http://bitcoin.org/bitcoin.pdf. Accessed 2017

17. Lin, Q., Chang, P., Chen, G., Ooi, B.C., Tan, K., Wang, Z.: Towards a non-2PC transaction management in distributed database systems. In: Proceedings ACM International Conference Management Data, pp. 1659–1674 (2016)

18. Thomson, A., Diamond, T., Weng, S., Ren, K., Shao, P., Abadi, D.J.: Calvin: fast distributed transactions for partitioned database systems. In: Proceedings ACM International Conference Management Data, pp. 1–12 (2012)

19. Bailis, P., Fekete, A., Franklin, M.J., Ghodsi, A., Hellerstein, J.M., Stoica, I.: Coordination avoidance in database systems. Proc. VLDB Endowment **8**(3), 185–196 (2014)

20. Vukolic, M.: The quest for scalable blockchain fabric: proof-ofwork versus BFT replication. In: Proceedings of International Workshop Open Problems Network Security - iNetSec, pp. 112–125 (2015)

Ultrasound-Coupled Electrocoagulation Based Azo Dye Fading Rate Prediction Using Deep Neural Networks

Meryem Akoulih[1], Smail Tigani[2(✉)], Asmaa Wakrim[1], Abdellah Chehri[4],
Rachid Saadane[3], and Sanae El Ghachtouli[1]

[1] IME Lab,Faculty of Science, Hassan 2 University, 20250 Casablanca, Morocco
[2] AAIR Lab, Digital Engineering and Artificial Intelligence Systems High Private
School, 20100 Casablanca, Morocco
s.tigani@epsinsia.com
[3] Electrical Engineering Department, SIRC-LaGeS, Hassania School of Public
Labors, 20250 Casablanca, Morocco
[4] Department of Mathematics and Computer Science, Royal Military College,
Kingston, ON K7K 7B4, Canada

Abstract. This research paper presents a deep learning based predictive model for predicting the azo dye fading rate during electrocoagulation coupled with ultrasound. Electrocoagulation is a commonly used method for the treatment of wastewater contaminated with azo dyes, but the efficiency of this process can be improved by incorporating ultrasound. The proposed deep learning model is trained on a dataset of electrocoagulation experiments with varying parameters such as current density, pH, and treatment time, as well as the presence or absence of ultrasound. The model is able to accurately predict the azo dye removal rate for new sets of electrocoagulation parameters, providing a valuable tool for optimizing the treatment process. The model's performance is evaluated using standard metrics, and the results demonstrate its effectiveness in predicting azo dye removal rate.

Keywords: Electrocoagulation · Ultrasound · Wastewater · Azo dye · Deep learning

1 Introduction

Dyes are a class of organic compounds that are widely used in various industries for their ability to impart color to a wide range of materials, including textiles, paper, plastics, and food. One of the most widely used classes of dyes is the azo dyes, which are characterized by the presence of the azo (-N=N-) functional group in their chemical structure. Azo dyes are particularly important in the textile industry, where they are used to color fabrics and yarns. However, the widespread use of azo dyes has led to significant environmental concerns due

© The Author(s), under exclusive license to Springer Nature Singapore Pte Ltd. 2023
A. Zimmermann et al. (Eds.): KES-HCIS 2023, SIST 359, pp. 24–32, 2023.
https://doi.org/10.1007/978-981-99-3424-9_3

to their persistent nature and the potential for the release of toxic byproducts during their degradation.

The textile industry is one of the largest users of dyes, and azo dyes represent a significant portion of the dyes used in this industry. They are particularly popular due to their wide range of colors, excellent lightfastness, and good washing fastness. However, the use of azo dyes in the textile industry has led to a number of environmental concerns. The primary concern is that azo dyes are not easily biodegradable and can persist in the environment for long periods of time. This can lead to the accumulation of these compounds in soil and water, which can have negative impacts on the health of both humans and the environment.

Another concern is that azo dyes can be broken down by microorganisms in the environment to release potentially toxic byproducts. These byproducts, such as amines, can be harmful to aquatic life and can also have negative impacts on human health if they are present in drinking water. Additionally, azo dyes can also be toxic to the microorganisms responsible for their degradation, which can disrupt the natural balance of ecosystems.

The release of azo dyes into the environment is of particular concern because they are used in such large quantities in the textile industry. Wastewater from textile factories can contain high levels of azo dyes, and this can have a negative impact on the quality of nearby water bodies. The treatment of textile wastewater is therefore of critical importance for the protection of the environment.

In light of these environmental concerns, researchers have been working to develop more sustainable methods for the treatment of textile wastewater. One area of focus has been the use of electrocoagulation in combination with ultrasound for the treatment of azo dye-containing wastewater. Electrocoagulation is a process that uses electrical current to form flocs (clumps) of particles in water, which can then be separated out and removed [7]. Ultrasound is a mechanical energy that is used to enhance the flocculation process. This research aims to develop a deep learning predictive model, as in [8], to predict the fading rate of azo dyes in textile wastewater using electrocoagulation coupled with ultrasound [6].

2   Materials and Methods

2.1   Electrocoagulation

Electrocoagulation is a technology that uses electrical current to promote the formation of flocs, or clumps, of particles in water. These flocs can then be separated out and removed from the water, making it possible to remove a wide range of contaminants, including dyes. In the context of azo dyes, electrocoagulation can be used to promote the fading of the dyes as well as their removal from the water.

The basic principle of electrocoagulation is that an electrical current is applied to an electrolyte solution, typically water or wastewater, which causes the formation of hydroxide ions (OH-) and hydrogen gas (H2) at the anode and the release of hydroxyl ions (OH+) and oxygen gas (O2) at the cathode. These hydroxide and hydroxyl ions react with the contaminants present in the water, such as dyes, to form flocs, which can then be removed by sedimentation, flotation or filtration [5].

The electrocoagulation process has several advantages when it comes to removing dyes from water. One of the main advantages is that it can be used to treat a wide range of dyes, including azo dyes. Additionally, electrocoagulation is relatively simple to operate and can be performed at ambient temperature, which reduces the energy requirements of the process [2]. Furthermore, the electrocoagulation process can also be easily integrated with other treatment technologies, such as adsorption and biodegradation, to enhance the removal efficiency of dyes.

2.2   Ultrasound

Ultrasound is a mechanical energy that is characterized by the generation of high-frequency sound waves [9]. These sound waves are at a frequency higher than the audible range for humans, typically above 20 kHz. Ultrasound can be used to enhance a wide range of chemical and physical processes, including the treatment of wastewater containing dyes.

In the context of azo dyes, ultrasound can be used to enhance the fading process as well as the removal of the dyes from the water. The mechanism behind this enhancement is not fully understood, but it is thought that the high-frequency sound waves create microscopic bubbles in the liquid that implode with great force. This implosion creates high-energy environments, such as high-pressure, high-temperature and high shear force, which can aid in the fading of the dyes and the formation of flocs.

The use of ultrasound in the treatment of dyes-containing wastewater has several advantages. Firstly, it can be used to enhance the performance of other treatment technologies, such as electrocoagulation, which can lead to higher removal rates of dyes. Additionally, ultrasound can be used to reduce the time and energy required for treatment. The use of ultrasound can also reduce the consumption of chemical reagents that are commonly used in the treatment of dyes-containing wastewater. Furthermore, ultrasound has the ability to remove dyes that are not easily removable by other methods, and it has been shown to be effective in removing recalcitrant dyes.

2.3   Ultrasound-Coupled with Electrocoagulation

Ultrasound coupled with electrocoagulation is a combined technology that uses the advantages of both ultrasound and electrocoagulation to enhance the fading and removal of azo dyes from water [4]. The basic principle of this technology is that an electrical current is applied to an electrolyte solution, typically water or wastewater, which causes the formation of hydroxide ions (OH-) and hydrogen gas (H2) at the anode and the release of hydroxyl ions (OH+) and oxygen gas (O2) at the cathode.

At the same time, high-frequency sound waves are applied to the solution, creating microscopic bubbles that implode with great force, creating high-energy environments, such as high-pressure, high-temperature and high shear force [1]. The combination of the high-energy environments created by the ultrasonic waves and the hydroxide and hydroxyl ions released by the electrocoagulation process, enhances the fading and removal of the azo dyes.

The use of ultrasound coupled with electrocoagulation has several advantages over the use of each individual technology. Firstly, this combined technology can lead to higher removal rates of dyes than either technology alone. Additionally, the use of ultrasound coupled with electrocoagulation can reduce the time and energy required for treatment, as well as the consumption of chemical reagents. Furthermore, this combined technology has the ability to remove dyes that are not easily removable by other methods, and it has been shown to be effective in removing recalcitrant dyes, which are dyes that are hard to remove by other methods.

3   Data Analytics

3.1   Data Set Overview

Table 1 shows the experimental results obtained using Electrocoagulation combined with ultrasound. The input parameters P_{CT} as the exercise time, P_I as the applied electric intensity, P_{pH} as the applied pH, C_{NaCl} as the concentration and P_T as the temperature. Let E_{FR} be the experimentally observed decay rate.

3.2   Data Set Descriptive Statistics

This subsection focuses on the description of the input and output parameters. We compute mainly - for each parameter - the minimum observed value, the maximum, the mean and the standard deviation. Table 1 describes the input data :

Table 1. Input Configuration Descriptive Statistics

	P_{CT}	P_I	P_{pH}	C_{NaCl}	P_T
Min	5	1.0	3	0	25
Max	40	5.0	10	200	60
Mean	25.42	4.48	7.75	29.17	29.58
Standard Deviation	12.39	1.22	1.54	59.68	10.75

Table 2 describes the fading rate seen as the output observed value :

Table 2. Fading Results Descriptive Statistics

	Fading Rate E_{FR}
Min	7.44
Max	99.52
Mean	78.32
Standard Deviation	26.95

3.3 Pearson Correlation

Karl Pearson's correlation coefficient in Table 3 can be used to summarize the strength of the linear relationship between two data samples. Pearson's correlation coefficient is computed as the covariance of the two variables divided by the product of the standard deviations of each data sample. This is his covariance normalization between two variables to get an interpretable score. It is given formally with the equation :

$$C_{XY} = \frac{\sum_{i=1}^{n}(x_i - \bar{X})(y_i - \bar{Y})}{\sqrt{\sum_{i=1}^{n}(x_i - \bar{X})^2} . \sqrt{\sum_{i=1}^{n}(y_i - \bar{Y})^2}} \tag{1}$$

In this case E_{FR} that represents the experimental fading rate (X) and the (Y) variables represent each input parameter of the setup described in the previous subsection. P_{CT} experience time, P_I applied electric intensity, P_{pH} applied pH value, C_{NaCl} concentration and P_T P_{CT} are is the stress time, P_I is the applied electric intensity, P_{pH} is the applied pH value, C_{NaCl} is the concentration, and P_T.

Using the Eq. 1 produces a Pearson's correlation coefficient and a p-value to test for lack of correlation. Using the mean and standard deviation in the calculation, we know that the two data samples should have a Gaussian or Gaussian-like distribution.

Table 3. Pearson Correlation between Fading Rate and Input Parameters

	Pearson Coefficient	p-Value
Fading/Time	0.100324	0.330777
Fading/Intensity	0.528927	0.0
Fading/pH	−0.079325	0.442338
Fading/NaCl	0.313505	0.001869
Fading/Temperature	0.240131	0.018445

The resulting correlation coefficient can be interpreted to understand the relationship. The coefficient returns a value between −1 and 1, representing the limits of correlation from completely negative correlation to completely positive correlation. A value of 0 means no correlation. You have to interpret the value.

Values below -0.5 or above 0.5 often show significant correlation, and values below these values show less significant correlation. The p-value roughly indicates the probability that an uncorrelated system will produce datasets with at least as extreme Pearson correlations as those calculated from those datasets.

3.4 Deep Neural Network

Table 4 shows the obtained predictions using our deep learning model compared to experimental observed fading rates. The deep learning predictive model proposed in this research was trained using a dataset of azo dye fading rates obtained from experiments using electrocoagulation coupled with ultrasound. The training process involved the use of a supervised learning algorithm, where the model was trained to predict the fading rate of azo dyes based on a set of input parameters. The training process was carried out using a software library that was specifically designed for deep learning: TensorFlow in our case.

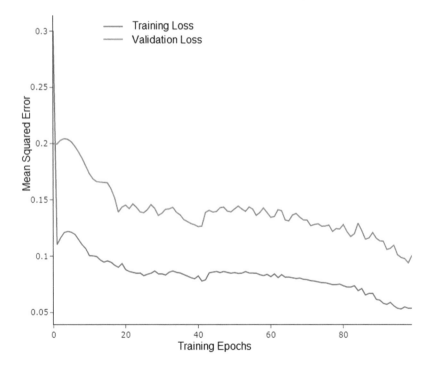

Fig. 1. Deep Neural Network Training and Validation Loss Evolution

The figure below illustrates the evolution of the loss function during the training and validation phases of the model. The loss function is a measure of how well the model is able to predict the fading rate of azo dyes, and it is calculated as the difference between the predicted fading rate and the actual

Table 4. Fading Rate Predictions using Trained Deep Learning Model

P_{CT}	P_I	P_{pH}	C_{NaCl}	P_T	E_{FR}	$\bar{E}_{FR}$	ε
20	5	8	50	25	0.9873	0.9841	0.0032
15	1	8	0	25	0.2150	0.2391	−0.0240
15	5	8	200	25	0.9912	0.9724	0.0188
15	5	3	0	25	0.9550	0.9544	0.0014
30	5	8	0	25	0.9796	0.9815	−0.0018
25	5	8	0	25	0.9769	0.9769	0.0000
25	5	8	200	25	0.9927	0.9955	−0.0028
10	5	8	0	60	0.8868	0.8879	−0.0011
35	5	8	50	25	0.9922	0.9913	0.0009
15	5	8	0	60	0.9893	0.9736	0.0157
40	5	8	0	60	0.9952	0.9963	−0.0011
10	5	8	0	45	0.8076	0.8155	−0.0079
40	2.8	8	0	25	0.9494	0.9492	0.0002
35	1	8	0	25	0.4575	0.4751	−0.0176
30	5	8	0	60	0.9908	0.9909	−0.0001
20	5	10	0	25	0.9676	0.9649	0.0027
30	5	8	0	45	0.9918	0.9892	0.0026
30	1	8	0	25	0.3880	0.4117	−0.0237
25	1	8	0	25	0.3564	0.3746	−0.0181
25	5	8	50	25	0.9906	0.9885	0.0021
5	5	8	0	25	0.6108	0.6100	0.0009
30	5	8	200	25	0.9896	0.9978	−0.0082
25	5	8	0	45	0.9815	0.9807	0.0008
10	5	8	200	25	0.9526	0.8961	0.0565
5	5	10	0	25	0.4823	0.4943	−0.0120
5	5	8	0	60	0.7056	0.7072	−0.0016
10	5	8	50	25	0.8705	0.8634	0.0071
40	5	8	0	25	0.9838	0.9861	−0.0022
10	5	10	0	25	0.6315	0.6433	−0.0118
25	2.8	8	0	25	0.9005	0.9199	−0.0193
25	5	8	0	60	0.9896	0.9885	0.0011
20	5	8	100	25	0.9942	0.9879	0.0063
20	2.8	8	0	25	0.7553	0.8065	−0.0512
10	5	8	100	25	0.9118	0.8914	0.0204
...	...	...	...	...	...	...	...

(*continued*)

Table 4. (*continued*)

P_{CT}	P_I	P_{pH}	C_{NaCl}	P_T	E_{FR}	$\bar{E}_{FR}$	ε
5	5	3	0	25	0.4115	0.4205	−0.0090
35	5	10	0	25	0.9819	0.9798	0.0022
30	5	8	50	25	0.9921	0.9902	0.0019
15	5	8	0	45	0.9930	0.9396	0.0534
10	2.8	8	0	25	0.3653	0.4682	−0.1029
5	2.8	8	0	25	0.2890	0.6909	−0.4019
15	5	8	0	25	0.9557	0.9093	0.0464
5	5	8	0	45	0.6000	0.6643	−0.0643
15	2.8	8	0	25	0.5932	0.6450	−0.0518
35	2.8	8	0	25	0.9376	0.9445	−0.0068
5	5	8	50	25	0.7130	0.6602	0.0528
40	5	3	0	25	0.9911	0.9931	−0.0019
20	5	8	0	45	0.9921	0.9607	0.0314
20	1	8	0	25	0.2364	0.3223	−0.0859
10	5	8	0	25	0.6834	0.7118	−0.0283
25	5	3	0	25	0.9867	0.9874	−0.0007
20	5	3	0	25	0.9819	0.9804	0.0015
30	5	10	0	25	0.9790	0.9763	0.0027

fading rate. The training phase is represented by the blue line, and the validation phase is represented by the orange line. As the training progresses, the loss function decreases, indicating that the model is becoming more accurate in its predictions. The validation loss is used to evaluate how well the model generalize to new unseen data. Table 4 reports the predictions that we obtained using our deep learning model. As explained before, the input parameters P_{CT} as the exercise time, P_I as the applied electric intensity, P_{pH} as the applied pH, C_{NaCl} as the concentration and P_T as the temperature. Let E_{FR} be the experimentally observed decay rate. In this subsection, we call $\bar{E}_{FR}$ the predicted fading rate and ε the prediction error that we obtain with the equation : $E_{FR} - \bar{E}_{FR}$.

4   Conclusion

In conclusion, this research has successfully demonstrated the potential of using a deep learning predictive model for the prediction of azo dye fading rate through the use of electrocoagulation coupled with ultrasound. The results show that the proposed model is able to accurately predict the fading rate of azo dyes in a highly efficient manner. The combination of electrocoagulation and ultrasound was found to greatly enhance the efficiency of the fading process, and the deep learning predictive model was able to effectively capture this enhancement. This research has significant implications for the treatment of textile wastewater in industries that use azo dyes, as it provides a more efficient and accurate method for predicting fading rates. Overall, the proposed model represents a valuable tool for the optimization of electrocoagulation-ultrasound processes for the treatment of azo dye-containing wastewater [3].

Acknowledgment. My gratitude to Pr. Amal Tazi for all what she have done, which I will never forget. I truly appreciate her and the time she spent helping me in many occasions. I would like to thank the anonymous referees for their valuable comments. Special thanks goes to any one that improved the language's quality of this paper.

References

1. Rossi, G., Mainardis, M., Aneggi, E., Weavers, L.K., Goi, D.: Combined ultrasound-ozone treatment for reutilization of primary effluent—a preliminary study. Environmental Science and Pollution Research **28**(1), 700–710 (2020). https://doi.org/10.1007/s11356-020-10467-y
2. Magnisali, E., Yan, Q., Vayenas, D.V.: Electrocoagulation as a revived wastewater treatment method-practical approaches: a review. J. Chem. Technol. Biotechnol. **97**(1), 9–25 (2022)
3. Karabacakoğlu, B., Tezakil, F.: Electrocoagulation of corrugated box industrial effluents and optimization by response surface methodology. Electrocatalysis **14**(2), 159–169 (2023) https://doi.org/10.1007/s12678-022-00781-z
4. Dolatabadi, M., Kheirieh, A., Yoosefian, M.: Hydroxyzine removal from the polluted aqueous solution using the hybrid treatment process of electrocoagulation and adsorption; optimization, and modeling. Appl. Water Sci. **12**(11), 254 (2022). https://doi.org/10.1007/s13201-022-01780-7
5. Shaker, O.A., Safwat, S.M., Matta, M.E.: Nickel removal from wastewater using electrocoagulation process with zinc electrodes under various operating conditions: performance investigation, mechanism exploration, and cost analysis. Environ. Sci. Pollut. Res. **30**, 26650–26662 (2023) https://doi.org/10.1007/s11356-022-24101-6
6. Arka, A., Dawit, C., Befekadu,A., Debela, S.K., Asaithambi, P.: Wastewater treatment using sono-electrocoagulation process: optimization through response surface methodology. Sustain. Water Resour. Manage. **8**(3), 61 (2022)
7. Ilhan, F. Ulucan-Altuntas, K. Avsar, Y. Kurt, U., Saral., A.: Electrocoagulation process for the treatment of metal-plating wastewater: kinetic modeling and energy consumption. Frontiers Environ. Sci. Eng. **13**(5), 73 (2019)
8. Lin, W., Hanyue, Y., Bin. L.: Prediction of wastewater treatment system based on deep learning. Frontiers Ecol. Evol. **10**, 1064555 (2022)
9. Son, Y.: Advanced oxidation processes using ultrasound technology for water and wastewater treatment, pp. 1–22. Springer, Singapore (2015). https://doi.org/10.1007/978-981-287-470-2_53-1

Digital Strategy and Architecture for Human-Centered Intelligent Systems

Alfred Zimmermann[1]([⊠]), Rainer Schmidt[2], Rainer Alt[3], Yoshimasa Masuda[4,5,6], and Abdellah Chehri[7]

[1] Reutlingen University Herman Hollerith Center, Böblingen, Germany
alfred.zimmermann@reutlingen-university.de
[2] Munich University of Applied Sciences, Munich, Germany
rainer.schmidt@hm.edu
[3] University of Leipzig, Leipzig, Germany
rainer.alt@uni-leipzig.de
[4] Tokyo University of Science, Tokyo, Japan
ymasuda@andrew.cmu.edu
[5] Keio University, Tokyo, Japan
[6] Carnegie Mellon University, Pittsburgh, USA
[7] Royal Military College of Canada, Kingston, Canada
abdellah.chehri@rmc-cmr.ca

Abstract. Current advances in Artificial Intelligence (AI) combined with other digitalization efforts are changing the role of technology in service ecosystems. Human-centered intelligent systems and services are the target of many current digitalization efforts and part of a massive digital transformation based on digital technologies. Artificial intelligence, in particular, is having a powerful impact on new opportunities for shared value creation and the development of smart service ecosystems. Motivated by experiences and observations from digitalization projects, this paper presents new methodological experiences from academia and practice on a joint view of digital strategy and architecture of intelligent service ecosystems and explores the impact of digitalization based on real case study results. Digital enterprise architecture models serve as an integral representation of business, information, and technology perspectives of intelligent service-based enterprise systems to support management and development. This paper focuses on the novel aspect of closely aligned digital strategy and architecture models for intelligent service ecosystems and highlights the fundamental business mechanism of AI-based value creation, the corresponding digital architecture, and management models. We present key strategy-oriented architecture model perspectives for intelligent systems.

Keywords: Intelligent Digitalization · Artificial Intelligence · Human-centered Intelligent Systems · Digital Strategy · Intelligent Service Architecture

© The Author(s), under exclusive license to Springer Nature Singapore Pte Ltd. 2023
A. Zimmermann et al. (Eds.): KES-HCIS 2023, SIST 359, pp. 33–42, 2023.
https://doi.org/10.1007/978-981-99-3424-9_4

1   Introduction

The current state of human-centric artificial intelligence [1] and visions for AI in 2041 [2] are fundamentally changing the role of technology for digital platforms [3] and human-centric intelligent systems [4, 5]. Digitization with smart service ecosystems as part of an open architectural environment often disrupts existing businesses, technologies, and economies. Today, digital transformation [6] is profoundly changing existing businesses and economies. The potential of the Internet and related digital technologies [4, 6], such as the Internet of Things, artificial intelligence, data analytics, services computing, cloud computing, mobile systems, collaboration networks, blockchains, cyber-physical systems, and Industry 4.0, are strategic drivers as well as enablers of digital platforms with rapidly evolving ecosystems of human-centered intelligent systems and services.

Influenced by the transition to digitalization, many companies are currently transforming their strategy, culture, processes, and information systems to drive digitalization and introduce artificial intelligence systems and services. Human-centric intelligent systems are information systems that use artificial intelligence (AI) [5] to support and interact with humans [11]. Current advances in artificial intelligence have led to a rapidly growing number of intelligent services and applications.

However, there is a lack of a concrete methodological framework for designing and linking a digital strategy with intelligent service architectures and the resulting human-centric intelligent systems. Therefore, our paper can be underpinned by the following research question:

What are integral models and frameworks on the convergence of digital strategy and architecture for designing human-centric intelligent systems?

First, we establish the basic context of intelligent digitization and digital transformation. We provide insights into human-centric intelligent systems based on a fundamental AI architecture. At its core, we provide the basic orientation for a reference model of an innovative intelligent service architecture and link it to our current digital strategy framework. In the final section, we conclude our research findings and provide an outlook for our future work.

2   Smart Digitalization

Digital transformation is currently the predominant form of business transformation [4, 6], with IT acting as both a technological enabler and a strategic driver. Digital technologies are the key strategic drivers for digitization as they change the way businesses are run and have the potential to disrupt existing business processes. SMACIT defines the strategic core of digital technologies in [4], with abbreviations for Social, Mobile, Analytics, Cloud, Internet of Things. From today's perspective, we need to expand this technological core to include artificial intelligence and cognition, biometrics, robotics, blockchain, 3D printing, and edge computing. Digital technologies deliver three core capabilities for a fundamentally changing business: ubiquitous data availability, unlimited connectivity, and massive computing power.

Initially, digitization was considered a primarily technical term [4]. As a result, a number of technologies are often associated with digitalization [7]: Cloud Computing,

Big Data combined with advanced analytics, Social Software, and the Internet of Things. New technologies such as Deep Learning are strategic enablers and closely linked to the advances of digitalization. They enable the use of computers for activities that were previously the exclusive domain of humans. Therefore, the current focus on intelligent digitalization is an important area of research. Digital services and related products are software intensive [4] and therefore adaptable and usually service oriented [8]. Digital products are able to extend their capabilities by accessing cloud services and change their current behavior.

When we use the term digitalization, we mean more than just digital technologies. Digitalization [9] bundles the more mature phase of a digital transformation from analog to digital to fully digital. Digital substitution (digitization) initially replaces only analog media with digital media, taking into account the same existing business values, while augmentation functionally enriches the corresponding transformed analog media. In a further step of digital transformation, new processing patterns or processes are made possible by digitally supported modification of the basic terms (concepts). Finally, the digital redefinition ("digitalization") of processes, services, products, processes and systems creates completely new forms of value propositions for disruptive companies, services, products and systems. Digitalization is therefore about shifting processes toward attractive highly automated digital business processes and not just about communication via the Internet. Digital redefinition usually has a disruptive impact on business. Beyond value-driven perspective of digitalization, intelligent systems require human, ethical and social orientations.

Considering the closely related concepts of digitization, digitalization and digital transformation [4] and [9], we conclude: Digitization and digitalization are about digital technology, while digital transformation is about the changing role of digital customers and the digital change process based on new value propositions. We digitize information, we digitize processes and roles for enhanced platform-based business operations, and we digitally transform [6] the enterprise by driving digital strategy, customer-centric and value-driven digital business models, and architecture-driven digital transformation.

3   Human-Centered Intelligent Systems

From today's perspective, probably no digital technology is more exciting than artificial intelligence, which offers massive automation opportunities for intelligent digital systems and services. Artificial intelligence (AI) [5, 10, 12] is often used in conjunction with other digital technologies such as analytics, ubiquitous data, the Internet of Things, cloud computing, and unlimited connectivity. Fundamental capabilities of AI involve automatically generated solutions from previous useful cases and solution elements derived from causal knowledge structures [5] such as rules and ontologies, as well as learned solutions based on data analytics with machine learning and deep learning with neural networks.

The study of artificial intelligence from a human-centered perspective [10] requires a deep understanding of the role of human ethics, human values and habits, and practices and preferences in developing and interacting with intelligent systems. With the success of artificial intelligence, new concerns and challenges arise regarding the impact of these technologies on human life. These include questions about the security and trustworthiness of AI technologies in digital systems, the fairness and transparency of systems, and the impacts of AI on people and society.

To address the growing amount of sensor data and unstructured data, we adopt MIT's canonical AI architecture model [11] (Fig. 1), which includes data preparation and curation, specific ML and DL algorithms and frameworks, and systematic integration of human-centric cognitive services. The AI reference architecture includes modern infrastructures and features to support robust AI, e.g., explainable AI.

Symbolic AI [5], which was prevalent until the 1990s, uses a deductive, expert-based approach. By consulting one or more experts, knowledge is gathered in the form of rules and other explicit knowledge representations, such as Horn clauses. These rules are applied to facts that describe the problem to be solved. The solution to a problem is found by successively applying one or more rules using the mechanisms of an inference engine. An inference path can usually be traced backward and forward, providing transparency and rationality about the instantiated inference processes through "how" and "why" explanations. Symbolic AI has proven to be very effective for highly formalized problem spaces such as theorem proving. After the last wave of enthusiasm in the late 1980s, the focus of research shifted to other areas [5, 11, 12].

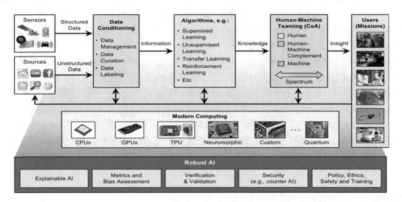

Fig. 1. Canonical AI Architecture [11].

Ontologies [5] represent the second wave of semantic technologies to support explicit knowledge representations. Ontologies have their background in the philosophy of being and existence. From the perspective of symbolic AI, ontologies are explicit, machine-readable representations of basic categories of concepts and their relationships. The Web Ontology Language OWL defines a family of knowledge representation languages for ontologies to represent the formal semantics of concepts and relationships with logical terms. Ontologies provide a common vocabulary for specific domains and, like rules, must be represented by manual design efforts.

From today's perspective, basic AI capabilities concern automatically learned solutions generated by data analysis with machine learning and deep learning, supported by neural networks and complementary methods. Neural networks [5, 10, 11] are inspired by the metaphor of the human brain, which connects artificial neurons via a network that receives input and produces output data. Together with genetic algorithms, fuzzy systems, rough sets, and the study of chaos, they are examples of new approaches to artificial intelligence. Deep learning is considered a subclass of machine learning approaches. Even more than machine learning, neural networks and deep learning are able to capture tacit knowledge. The fundamental mechanism of neural networks is the adaptation of weights representing the strength of connections between neurons until the conversion of input signals into output signals shows the desired behavior. The adaptation of weights using training data is called learning.

In contrast to symbolic AI, machine learning [11, 12] uses an inductive approach based on a large amount of analyzed data. We distinguish three basic approaches to machine learning: supervised, unsupervised learning, and reinforcement learning. In supervised machine learning approaches, the target value is part of the training data and is based on sample inputs. Typically, unsupervised learning is used to discover new hidden patterns within the analyzed data. Reinforcement Learning (RL) is an area of machine learning with software agents [5] working to maximize cumulative rewards. The exploration environment is specified in terms of a Markov decision process because many reinforcement learning algorithms use dynamic programming techniques.

AI-human compatibility [1, 10] refers to the ability of artificial intelligence (AI) systems to effectively interact and communicate with human users. This can include the ability to understand natural language, respond in a way that is easily understood by humans, and adapt to the user's preferences and needs. It also includes the ability to recognize and respond appropriately to human emotions and social cues. There are many different approaches to making AI more human-friendly, such as natural language processing, machine learning, and human-computer interaction research.

The combination of hardware and software product components with cloud-based intelligent services enables new perspectives for AI-based assistance platforms [13]. One example is Amazon Alexa, which combines a physical device with microphone and speaker with services called Alexa Skills. Users can extend Alexa's capabilities with Skills, which function similarly to apps. The set of Alexa Skills is dynamic and can be tailored to the customer's needs on the fly. Alexa enables voice interaction, music playback, to-do lists, setting alarms, streaming podcasts, playing audiobooks and providing weather, traffic, sports and other real-time information such as news.

4 Digital Strategy

Digital technologies are changing the way we communicate and collaborate to create value with customers and other stakeholders, and even competitors. Digital technologies have changed our view of how we can analyze and understand a variety of data accessible in real time from different perspectives. Digital transformation has also changed our understanding of how to innovate in global processes to design and develop smart digital products and services faster than ever before with the best available digital technology and quality.

Digitalization forces us to look differently at value creation for and together with customers and other co-creators or beneficiaries. While digital technologies are key strategic enablers [4] for new customer-centric digital businesses, there is a need to solve real customer problems and drive customer engagement by creating new value propositions, encouraging customer co-creation, and offering new digital solutions and services. To increase customer value, smart digital services and systems integrate digital and non-digital interactions. Digital systems and interfaces enable customer experiences across digital channels and improved service-oriented product features.

Major strategic trends [4] of our time represent fundamental changes for the next digital business, such as: Digitalization of products and services according to the service-dominant logic, contextual value creation, consumerization of IT, digitalization of work, digitalization of business models, especially to today's digital platforms and smart service ecosystems.

A digital strategy [14] is a combination of initiatives where a company selects online activities to achieve its business goals/vision. A successful digital strategy will both engage end users and build the business case for execution. Our models for optimizing a digital strategy even for short-term orientations come from [14] and [15]. Our strategy framework helps us develop a successful digital strategy and digital IT architecture with key orientation processes and recommended steps. The original framework (Fig. 2) groups the following ten essential aspects and refers to successful methods from the state of the art and practice: Strategic Context Analysis, Strategic Drivers, Digital Mission and Values, Strategic Scenarios, Digital Vision, Strategic Goals (Focus Areas), Strategic Objectives and BSC (Balanced Score Card) with KPIs, Mapping: Strategic Value for a Digital Operating Model, Strategic Initiatives (Projects), Strategic Portfolio and Roadmap, Integration of Digital Strategy with Tactics and Controlling.

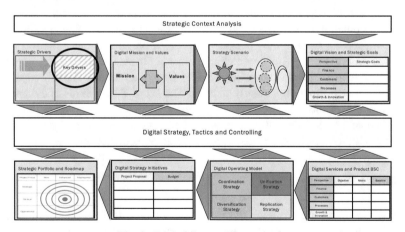

Fig. 2. Digital Strategy Framework.

The central point and main direction of a fundamental digital strategy is the strategic vision, which is aligned with the strategic goals you want to achieve. The vision is the anchor that keeps you from getting lost at sea. The vision directs the strategy toward the outcomes that matter most to an organization. The elements of the strategic plan will help you get closer to the digital vision. The vision should provide direction and inspiration to employees and help attract talent and investment to an organization.

By comprehensively combining AI technologies, algorithms, and strategic application areas in a new way, Kai-Fu Lee and Chen Qiufan have compiled ten visions for the AI future [2]. Strategic application areas of AI as part of this cluster include Big Data Analytics, AI in financial applications, computer vision, generative adversarial networks, biometrics, AI security, natural language processing, healthcare, robotics, virtual reality, augmented reality, brain-computer interface, ethical and social issues, autonomous vehicles, smart cities, quantum computing, autonomous weapons and existential threats, job displacement by AI, universal income and what AI can't do, happiness analytics, data analytics, privacy regulations, personal data, privacy computing, new economic models, and the future of money.

First, we model digital strategy [8, 14, 16] using the Business Model Canvas [18] and the Value Proposition Canvas [19], which provides direction for digital modeling and sets the foundation and value-based framework for the business definition models. With the basic models for a value-based digital enterprise, we map these basic service and product models to a digital operating model [4]. The value perspective of the Business Model Canvas [18] yields appropriate mappings to the enterprise architecture value models [20], which are in modern cases supported by ArchiMate modeling. Finally, we set the framework for digital services and associated products through the architecture of digital product compositions.

5   Intelligent Service Architecture

Digitalization [4, 6] fosters the development of IT systems with many, globally available and diverse, rather small and distributed cooperating structures, such as the Internet of Things or mobile systems. This has a strong impact on the architecture of intelligent digital services and products that integrate highly distributed intelligent systems. According to [15, 16] a service ecosystem is a self-contained, self-adaptive system of loosely coupled recourses that integrates actors connected by shared value creation through service exchange. In our understanding, a successful digital service platform [7, 15] should support a network of actors and host a set of loosely coupled open services and software products as part of a fast scalable digital ecosystem [8].

The Digital Enterprise Architecture Reference Cube (DEA) in Fig. 3 extends our holistic architecture reference and classification framework from [8] to drive bottom-up integration of dynamically composed microgranular architecture services and their models. The DEA cube encompasses governance, management and strategy perspectives as well as aligned core perspectives of a digital architecture. DEA abstracts from a particular business scenario or technology as it can be applied to different architectural instantiations to support smart ecosystems independent of different domains. DEA is complemented by governance methods from AIDAF [17].

DEA addresses platform and ecosystem architecture [3, 7, 8]. A digital platform is a repository of business, data, and infrastructure services used to rapidly configure digital offerings from digital services. Digital services and components are snippets of code that perform a specific task. We position reusable digital services as parts of an ecosystem of services. Furthermore, a digital platform linearizes the complexity of cooperating services. The value of a platform to users is derived from the number of users of platforms and services.

A blockchain platform [21] comprises a distributed software system that enables the creation and management of decentralized digital ledgers that can be used to log transactions over a computer network. These transactions are recorded in blocks that are chronologically linked together to form a chain, hence the name "blockchain." These platforms can be used for a variety of purposes, including digital currency transactions, smart contract execution, and supply chain management.

The business & information architecture [8, 20] combines business strategy with model structures for business products, business services, business control information, business domains, business process models, and business rules to create a specification framework for associated service-oriented information systems.

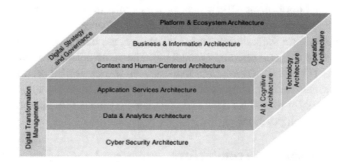

Fig. 3. Digital Enterprise Architecture Reference Cube (DEA Cube) [8].

According to Dey [22], a digital context includes all information that characterizes the situation of an entity and relates to an interaction between users, applications and the environment. An information needs context model is an extract from an enterprise model for a specific role that takes into account all the tasks of the role and is linked to the specific resources. A context acts like a set of constraints that influence the behavior of a system, a user or a computer. An information demand context model is an extract from an enterprise model for a specific role, taking into account all the roles' tasks and linked to the specific resources.

The AI and cognitive architecture outline basic components, services, mechanisms, and methods to support intelligent behavior of evolved digital systems and services. Cognitive computation [1] is inspired by neurobiology, cognitive psychology, AI, and connectionism. Connectionism represents a cognitive theory executed by adaptive and learning neural networks. Cognitive systems continuously build knowledge through learning, understand natural language, support problem solving, and interact with humans in more natural ways than conventionally programmed systems. An important aspect of AI technology is robust or trustworthy AI. This includes ways to explain AI results (e.g., why a system recommends a certain course of action), metrics to measure the effectiveness of an AI algorithm, verification and validation of intelligent systems, computer security, and the legal aspects that govern the safe, responsible, and ethical use of AI technology.

Data architecture describes and classifies the data structures used by an organization and its computer application software. Data architectures deal with stored, used, and moving data; descriptions of data stores, data groups, and data elements; and the mapping of data artifacts to qualities, applications, and locations.

The application services architecture [8, 20] summarizes the major application-specific service types and defines their relationship through a layered model of services that build on each other. The core functionality of the domain services is linked to the application interaction services and to the business process services of the customer organization. The core functionality of the domain services is linked to the application interaction capabilities and to the business processes of the customer organization.

The cybersecurity architecture specifies the organizational structure, standards, policies, and functional behavior of a computer network, including security and network functions. The cybersecurity architecture also describes how the various components of a cyber or computer system are organized, synchronized, and integrated. A cybersecurity architecture framework is a component of a system's overall architecture. It is developed and built to guide the development of an entire product/system.

Finally, the technology architecture [8] provides domain-independent software and hardware platforms to support the delivery of business, data, and application services. This includes IT infrastructures, platforms, middleware, networks, communications, processing, and related standards. The mapped operational architecture relies on service management processes to enable ongoing support and management of AI-based infrastructures of a digital enterprise.

6 Conclusion

We first outlined the context of intelligent digitization before focusing on human-centric intelligent systems by adopting MIT's canonical AI architecture. The main core of the paper comes from the integration of two methods by mapping our digital strategy framework together with the reference model of digital enterprise architecture as our basis for intelligent service architecture for human-centric intelligent systems. Based on our methods, we have answered the research question by presenting holistic, integral perspectives of the DEA - Digital Enterprise Architecture Reference Cube for the convergence of digital strategy and architecture structures to support the development of human-centered intelligent systems.

The strengths of our research stem from our novel approach to supporting intelligent digitization in the architecture of intelligent service ecosystems through an AI-powered co-creation model, an integral and scalable digital architecture reference model together with the framework for adaptive strategy engineering and management. Over the past 15 years, we have evaluated and optimized our original digital strategy and architecture models for intelligent service ecosystems by applying our methodologies and frameworks to real-world businesses and industry projects, and strategic innovation initiatives, as well as to a large number of academic case studies and student projects. Limitations of our work arise from the ongoing validation of our research and the open questions of the extended AI approaches, like generative AI, Explainable AI and associated inconsistencies between deep learning models and semantic dependencies.

Future research will address mechanisms, reference architectures, methods, and guidelines for a flexible and adaptive integration of digital strategies with architectures for AI-based service ecosystems.

References

1. Russel, S.: Human Compatible. Penguin Books, Artificial Intelligence and the Problem of Control (2019)
2. Lee, K.F., Qiufan, C.: AI 2041. Random House, Ten Visions for our Future (2021)
3. Hein, A., Weking, J., Schreieck, M., Wiesche, M., Böhm, M. and Krcmar, H.: Value Co-Creation Practices in Business-to-Business Platform Ecosystems. Electronic Markets, Vol. 29 No. 3, pp. 503–518, Springer (2019). https://doi.org/10.1007/s12525-019-00337-y
4. Ross, J.W., Beath, C.M., Mocker, M.: Designed for Digital. The MIT Press, How to Architect Your Business for Sustained Success (2019)
5. Russel, S., Norvig, P.: Artificial Intelligence: A Modern Approach. Pearson, London (2015)
6. Rogers, D.L.: The Digital Transformation Playbook. Columbia University Press, New York (2016)
7. McAfee, A., Brynjolfsson, E.: Machine, Platform, Crowd. Harnessing Our Digital Future. W. W. Norton & Company (2017)
8. Zimmermann, A., Schmidt, R., Sandkuhl, K., Masuda, Y., Chehri, A.: Architecting intelligent service ecosystems: perspectives, frameworks, and practices. In: Buchmann, R.A., Polini, A., Johansson, B., Karagiannis, D. (eds.) BIR 2021. LNBIP, vol. 430, pp. 150–164. Springer, Cham (2021). https://doi.org/10.1007/978-3-030-87205-2_10
9. Hamilton, E. R., Rosenberg, J. M., Akcaoglu, M.: Examining the Substitution Augmentation Modification Redefinition (SAMR) Model for Technology Integration. Tech Trends, 60, 433–441, Springer (2016)
10. Auenhammer, J.: Human-centered AI: the role of human-centered design research in the development of AI. In: Boess, S., Cheumg, M., Cain, R. (eds.) Synergy – DRS International Conference (2020)
11. Gadepally, V.N., et al: AI enabling technologies: a survey. MIT Lincoln Laboratory, ArXiv, abs/1905.03592 (2019)
12. Hwang, K.: Cloud Computing for Machine Learning and Cognitive Applications. The MIT Press, Cambridge (2017)
13. Schmidt, R., Alt, R., Zimmermann, A.: Ecosystem Intelligence for AI-based Assistant Platforms. HICSS (2022)
14. Bones, C., Hammersley, J., Shaw, N.: Optimizing Digital Strategy - How to Make Informed. Kogan Page, Tactical Decisions That Deliver Growth (2019)
15. Alstyne, M.W.V., Parker, G.G., Choudary, S.P.: Pipelines, platforms, and the new rules of strategy. Harv. Bus. Rev. **94**, 54–62 (2016)
16. Hagiu, A.: Strategic decisions for multisided platforms. MIT Sloan Manag. Rev. **55**, 71–80 (2014)
17. Masuda, Y., Viswanathan, M.: Enterprise Architecture for Global Companies in a Digital IT Era: Adaptive Integrated Digital Architecture Framework (AIDAF). Springer, Singapore (2019). https://doi.org/10.1007/978-981-13-1083-6
18. Osterwalder, A., Pigneur, Y.: Business Model Generation. John Wiley, Hoboken (2010)
19. Osterwalder, A., Pigneur, Y., Bernarda, G., Smith, A., Papadokos, T.: Value Proposition Design. John Wiley, Hoboken (2014)
20. Lankhorst, M.: Enterprise Architecture at Work. Springer, Berlin, Heidelberg (2017). https://doi.org/10.1007/978-3-662-53933-0
21. Nanayakkara, S., Rodrigo, M.N.N., Perera, S., Weerasuriya, G.T., Hijazi, A.A.: A methodology for selection of a Blockchain platform to develop an enterprise system. J. Ind. Inf. Integr. **23**. Elsevier (2021)
22. Dey, A.K.: Understanding and Using Context. Pers. Ubiquit. Comput. **5**, 4–7 (2001)

Smart Energy Management System: Methodology for Open-Pit Mine Power Grid Monitoring Applications

Younes Lemdaoui[1]([✉]), Adila Elmaghraoui[2], Mohamed El Aroussi[1], Rachid Saadane[1], and Abdellah Chehri[3]

[1] SIRC-(LaGeS), Hassania School of Public Works, Casablanca, Morocco
{ledmaoui.younes.cedoc,saadane}@ehtp.ac.ma
[2] Green Tech Institute (GTI), Mohammed VI Polytechnic University Benguerir, Benguerir, Morocco
adila.elmaghraoui@um6p.ma
[3] Department of Mathematics and Computer Science, Royal Military College of Canada, 11 K7K 7B4, ON Kingston, Canada
chehri@rmc.ca

Abstract. This paper proposes, describes, implements, and tests the Energy Monitoring System (EMS), a concept in data acquisition and transmission systems (DATS) applied to real-time cloud monitoring of a decentralized system in Industry 4.0. To arrive at this latest design, we went through various system projects. ISO 50001 is the most important. Energy monitoring is a critical component of project success. Our endeavor began in Benguerir's experimental open pit mine (OPM). It is built with three third party features: a NodeJS server, an EJS-based display, and a PostgreSQL database. Ethernet connectivity ensures data integrity and secrecy. As a result, our visualization includes current, voltage, power, energy, frequency, and power factor information.

Keywords: Energy Management · Industry 4.0 · Energy Monitoring · ISO 50001 · Smart Grid · Smart Meter

1 Introduction

The explosive growth of urbanization has presented the power industry with a formidable obstacle in the form of a significant challenge to the control of energy consumption, especially within the mining business.

The power sector will be in charge of managing the demand from customers. One of the responses to the demand-side energy management problem proposed in recent years is the design of energy management systems. Recently, these solutions have been provided in large numbers [1]. The EMS, using various technologies, works together, such as smart grids, microgrids, and smart cities. Local area networks are utilized here to control consumer goods so that we can guarantee the safety of our data (LANs).

© The Author(s), under exclusive license to Springer Nature Singapore Pte Ltd. 2023
A. Zimmermann et al. (Eds.): KES-HCIS 2023, SIST 359, pp. 43–53, 2023.
https://doi.org/10.1007/978-981-99-3424-9_5

The project was implemented in the experimental open-pit mine of Benguerir and is based on the architecture of an ethernet network in order to improve the data's level of security and reduce our reliance on the internet [2].

The system engineering approach is used during the design phase, whereas the MVC model is utilized during the development phase (Model–View–Controller). As a result, our endeavor will be utilized during the monitoring phase of the Deming wheel included in the ISO 50001 standard [3].

The purpose of this study is to achieve a secure solution in mining industries that consume a lot of energy and, as a result, spend a lot of money as a result of penalties either for an overwritten power or a bad power factor and as a consequence add a layer of supervision that adapts to the ISO 50001 standard [4].

The challenges of collecting, transmitting, and saving massive amounts of data in energy running processes can be managed by using various methods, such as digital instruments, communication network, software, database, and so on, and an energy management system based on Ethernet connectivity. To sum up:

- Customers are unable to monitor their daily energy usage because there is no apparatus or method available to indicate consumption that must be paid for. This is one of the problems.
- Every single one of our clients wants to cut their energy consumption by reducing the time spent using particular appliances.
- To build a distribution board system capable of managing energy by adding sensors. This is one of the most important goals of the proposed project.
- To save real-time data on energy use in the central processor every five minutes and to alert customers in the local processor every sixty minutes.
- Extraction of CSV files, including daily consumption and power factor profiles The following describes the format of this paper: Sect. 2 provides a brief overview of related works. Section 3 gives a description of the methodology and the metrics used to measure energy efficiency. In Sect. 4, we discuss the architecture that has been suggested. The results and discussion are given in Sect. 5. The conclusion and the article's perspective could be addressed in Sect. 6.

.

2   Related Works

Managing the demand for electrical energy brings together strategies to lower the amount of energy consumed by the industry, which results in cost savings (cost control). An electrical energy management system contributes to the protection of the environment. We have numerous ways that individuals can fight against the waste of energy, and one of those ways is an electrical energy management system.

For example, the reduction is achieved by using variable-speed drives for motors, pumping systems, and air compressors.

If the subscribed power is exceeded, the load shedder will stop the operation of any circuits classified as having a low priority. This pausing or idling only occurs when the total power drawn throughout the measurement period threatens to exceed the threshold for the fixed power limit.

The order of priority for offloading the equipment was previously determined and saved in the PLC. This order is followed when the equipment is unloaded. For instance, priority levels will be assigned so that the device whose interruption is most likely to be felt will be interrupted last. This will ensure that the least amount of disruption is caused.

"Reloading" refers to the process that occurs as demand decreases and an available power reserve is reconstructed. With monitoring technologies, keep a live eye on your consumption to have a better handle on it.

At this point, much progress has already been made within the scope of oversight. Every time, the manufacturers have continued to develop the functionality of their products in accordance with the demands placed on them by the users.

All of the home automation systems that are currently available are different from one another due to the fact that the operational context primarily determines the technology and the requirements of the customers. Because this is the social environment in which we find ourselves, we have been exposed to various works carried out on the management of electrical energy, alarm systems, and remote controls. The following are some of the works that captured our attention and are presented below:

2.1   The Design of an Internet of Things-Based Energy Monitoring System

This mode of data transmission remains vulnerable and is not secure, and the link is between the panels via the inverter and the network [4]. Even though this work was functional, it had three significant limitations, which are as follows: the use of the internet to send data from the sensors; this mode of data transmission remains vulnerable and is not secure; The functionality for processing and extracting excel files is not present.

2.2   The Role of Strategic Energy Management in the Environment of Industry 4.0

Because of the functional limitations imposed by this architecture on using TIEM as a PHP visualization tool, this architecture uses two distinct programming languages: JavaScript with the Nodejs framework on the server side and PHP with the Symfony framework on the view side. The architecture in question is not connected to the internet and uses a RabbitMQ connector to communicate with the MYSQL database [5].

For our project, we will document our attempts to contribute something of value to the work that has already been done in the monitoring of energy consumption in the mining industry; the experimental mine of Benguerir will serve as our point of reference for this endeavor. Following in the footsteps of the successes of the labor of our forebears, we are planning to install a module (system) that will make it possible to have real-time supervision of the posts on all of the electrical appliances in an effort to cut down on the amount of electricity that is consumed.

Our system will not only be restricted to lowering the amount of electricity that is consumed but it will also be required to account for the total amount of kilowatt-hours (kwh) of electrical energy that is consumed across the entire property using a three-phase digital electrical energy meter that is positioned upstream of the point where any circuit branches off. Nonetheless, its usage of electrical energy can be tracked. This ability to trace will serve as the basis for a decision. The goal of this project is to create a tool that will make it possible for people to adopt new attitudes or habits regarding consumption.

3   Methodology and Metrics for Measuring Energy Efficiency

The Energy Management System (EMS) has a few different goals in mind for its purposes. It is possible to attain and sustain optimal energy procurement and use across the company with a more capable EMS. This is something that can be accomplished. In addition, the EMS is able to reduce energy expenditures and waste without having an adverse effect on output or quality. Also, the EMS makes it possible to lessen reliance on imported goods while simultaneously improving energy security, economic competitiveness, and environmental quality.

Most importantly, it is anticipated that the EMS will make a substantial contribution to preserving the environment and the climate.

3.1   ISO 50001: Energy Management System

The International Standard for Energy Management Systems (EnMS) is referred to as ISO 50001 [6]. This standard is intended to assist companies in establishing energy management systems (EnMS) that are both efficient and effective, as well as in improving their energy performance. Companies that put these standards into practice have a better chance of lowering their energy efficiency costs, reducing their carbon dioxide emissions, and prioritizing the preservation and sustainable engagement with the environments in which they operate. These standards are based on the principles of continuous improvement, which have gained popularity as a result of the ISO 9001 and ISO 14001 management system standards [7].

An energy management system, often known as an EnSM, is the product of an ISO 50001 standard that has been effectively implemented. It is a term that describes everything that allows for the coordination of energy production and energy consumption and that is involved in the process of energy management.

3.2   Plan, Do, Check, Act (PDCA)

The Plan-Do-Check-Act (PDCA) structure for continuous improvement serves as the foundation for this International Standard, which integrates energy management into the day-to-day operations of a business.

Figure 1 depicts the numerous actions and steps that could arise during the PDCA process.

The following is an explanation of how the PDCA approach can be used in the context of energy management:

1. To Plan: to carry out the energy review and establish the baseline, energy performance indicators (EPIS), objectives, targets, and action plans required to deliver results that will improve

2. To Do: Put the action plans for energy management into effect.
3. To Check: Monitor and take measurements of the processes and the critical aspects of operations responsible for determining energy performance.
4. To Act: Take steps to enhance the organization's EnMS and energy performance consistently [8].

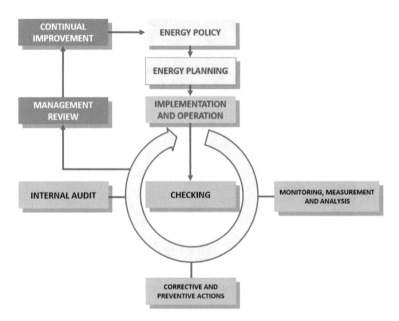

Fig. 1. Management System Model for iso 50001.

3.3 Adoption of a Standard Operating Procedure for the Energy Management System

The suggested system for managing energy consumption will be incorporated into the verification layer to determine the company's key performance indicators (KPIs). This will make it possible to take action on the processor and confirm the subsequent step. At certain intervals, the essential aspects of operations responsible for determining energy performance are monitored, measured, and assessed. This fascinating stage of the ISO 50001 standard makes it possible to:

1. Take corrective and preventative actions for significant energy consumption and other outputs.
2. Indices of Energy Efficiency and Performance (EPIs).
3. The efficiency of action plans.
4. A plan for measuring energy usage.
5. The analysis and rectification of any inconsistencies in the energy system's performance.
6. Tracking the records.

3.4 Metrics for Measuring Energy Efficiency

The Energy Management System allows for the definition of a number of metrics. As a result of the fact that these metrics contain measured values and ratios that determine the energy performance of the organization, they are able to serve as a reference against which any future changes in energy performance may be evaluated.

For example, the indicators of energy performance, also abbreviated as EPIs, are quantitative measurements of energy performance that are utilized to determine changes in energy consumption.

In our case, the EPIs are calculated and adjusted by maintaining a record of the energy usage for each month and dividing that figure by the total output charge. This process is repeated regularly. Figure 2 show the example of the proposed solution.

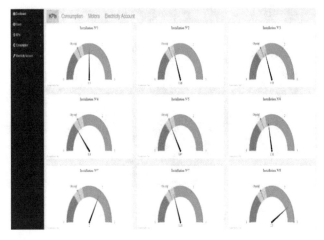

Fig. 2. The monitoring of EPIs (the proposed solution)

4 Material and Proposed Solution

4.1 System Structure

The Modbus Ethernet device driver collaborates with KEPServerEX to make it easier for OPC clients and PLCs to exchange data with one another when using a Modbus Ethernet protocol supported by the server. This occurs between the sensor and the server (as shown in Fig. 3). Our data collecting is automatically improved by KEPServerEX in response to the needs of the customers.

4.2   Sepam S40

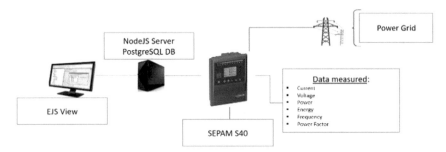

Fig. 3. The structure of the energy monitoring system.

As a result of the Sepam series 40's execution of the fundamental control and monitoring responsibilities required for the running of the electrical network, the requirement for auxiliary relays in the network is reduced. This is because the Sepam series 40 performs these duties.

Each Sepam comes preloaded with the command and control and monitoring features required for the task at hand. The Sepam S40 has three distinct varieties of user machine interfaces, four current inputs, three voltage inputs, and three analog inputs. In addition, ten logic inputs are available on the Sepam S40, and there are 88 relay outputs. Moreover, the logic equation editor, along with one port for communication through Modbus 16 inputs for temperature sensors are incorporated into the Sepam S40 module. This module is a part of the Sepam family (as shown in Fig. 4).

Fig. 4.   SEPAM S40.

4.3   Application

The following is a description of the overall architecture of our application when it is installed on a single machine. Figure 5 depicts the architecture as well as its individual modules.

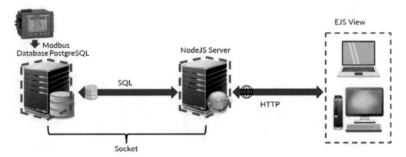

Fig. 5. The proposed architecture of the application.

1- **Connectivity with the OPC:** OPC is a widely recognized industrial communication standard that eliminates the need for proprietary constraints to facilitate the free data interchange between control applications and devices from several vendors. PLCs on the factory floor, RTUs out in the field, HMI stations and software applications running on desktop PCs can all participate in continuous data communication when using an OPC server. OPC compliance allows continuous real-time communication even when the underlying hardware and software are sourced from various manufacturers [9–11].

2- **NodeJS:** The "Node.js" is a JavaScript runtime environment that is open-source and allows code written in JavaScript to be executed outside of a browser. Developers can use JavaScript to construct command-line tools and for server-side scripting, which refers to running scripts on the server to produce dynamic web page content before the page is transmitted to the user's web browser. The "Node.js" makes this possible.

3- **EJS:** Embedded JavaScript, or EJS as its most often called. It is employed for the purpose of incorporating JavaScript code into HTML code. Express is the most common use for EJS within node.js. It performs the functions of a template engine and contributes to rendering JavaScript code on the client-side.

4- **Socket.IO:** A JavaScript library for use in developing real-time web applications is known as "Socket.IO". It makes it possible for web clients and servers to communicate with one another in real-time and in both directions. It consists of two components: a client-side library that can be launched in the browser and a server-side library that can be used with "node.js". Both elements share the same application programming interface.

5 Results and Discussions

The proposed system for monitoring energy consumption is purely web browser based. The primary data is stored in the form of an internal network that transmits in formation to the primary server. This allows for the mutual transmission and recording functions to be implemented, and it also makes it possible for the client to easily access the information stored on the primary server through their browser.

The web page that makes up the Web interface for the energy monitoring system can be modified to meet the needs of the design in accordance with the various criteria. The three following photos represent a different aspect of the significant web of energy monitoring systems (Fig. 6, Fig. 7 and Fig. 8].

Fig. 6. Authentication interface.

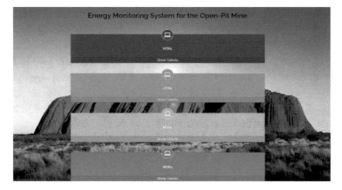

Fig. 7. Home interface.

Fig. 8. Measurements interface.

6   Conclusion and Perspectives

This article focuses primarily on the EMS as its primary topic of discussion. Throughout the design process, we used digital instrumentation, a communication network, and other technology. The technique is incredibly dependable and applicable in the software and database industries and other related technologies. The following is a list of some advantages of having an energy monitoring system: By contrasting and analyzing the management practices of various energy sources. The emergence of energy management systems is important in unified scheduling and optimization of enterprise energy, environmental quality improvement, lowering energy use, and increasing productivity. It also has a significant impact on the economy.

Energy management systems play a crucial role in enterprise energy's unified scheduling and optimization. The planning and carrying out of the accident, in addition to the root of the problem. Consequently, the proposed work gets one step closer to having a completely automated energy consumption monitoring system in which a computer can supply all of the essential information in addition to additional features based on its own assessment.

This monitoring tool in the mining sector has the potential to receive the following additional capabilities in the near future:

- Because of the prediction component, it will be feasible to create accurate projections on future consumption in relation to the load profile of the industry.
- After the chosen algorithm has been debated and tested, we will be able to integrate it into our application through an application programming interface (API) that facilitates communication between the client part of the application and the prediction server.
- In this context, three algorithms are planned for the experimentation phase following our database's data recorded in real-time.
- Finally, propose smart network architecture for underground mine monitoring. Based on application requirements and site surveys, it's important we develop a general architecture for this class of industrial applications [12]

References

1. Saleem, M.U., Usman, M.R., Shakir, M.: Design, implementation, and deployment of an iot based smart energy management system. IEEE Access **9**, 59649–59664 (2021). https://doi.org/10.1109/ACCESS.2021.3070960
2. Vishwakarma, S.K., Upadhyaya, P., Kumari, B., Mishra, A.K.: Smart energy efficient home automation system using IoT. In: 2019 4th International Conference on Internet of Things: Smart Innovation and Usages (IoT-SIU), pp. 1–4 (Apr 2019). doi: https://doi.org/10.1109/IoT-SIU.2019.8777607-
3. Laayati, O., Bouzi, M., Chebak, A.: Smart energy management system: design of a monitoring and peak load forecasting system for an experimental Open-Pit Mine. Appl. Syst. Innov. 5(1), Art. no. 1 (2022). doi: https://doi.org/10.3390/asi5010018
4. Luan, H., Leng, J. J.: Design of energy monitoring system based on IOT. In: 2016 Chinese Control and Decision Conference (CCDC), pp. 6785–6788 (May 2016). doi: https://doi.org/10.1109/CCDC.2016.7532219
5. Javied, T., Bakakeu, J., Gessinger, D., Franke, J.: Strategic energy management in indus- try 4.0 environment. In: 2018 Annual IEEE International Systems Conference (SysCon), pp. 1–4 (Apr 2018): https://doi.org/10.1109/SYSCON.2018.8369610
6. ISO 50001: Effective Energy Management Systems (EnMS), ISO Update. https://isoupdate.com/standards/iso-50001/ (Accessed 10 Feb 2023)
7. Taban, L.S.: ISO 50001: The Ultimate Guide to Energy Management Systems (EnMS) | Process Street | Checklist, Workflow and SOP Software (Sep. 30 2019). https://www.process.st/iso-50001/ (Accessed 10 Feb 2023)

8. ISO 50001, Default. https://www.e2singapore.gov.sg/overview/industry/energy-manage ment-systems/iso-50001 (Accessed 10 Feb 2023)
9. Systems Engineering, Default. https://www.incose.org/systems-engineering (Accessed 10 Feb 2023)
10. What is a Requirements Diagram - Knowledge Base, microTOOL. https://www.mi-crotool. de/en/knowledge-base/what-is-a-requirements-diagram/- (Accessed 10 Feb 2023)
11. What is an OPC Server | OPC Interoperability for KEPServerEX | Kepware. https://www.kep ware.com/en-us/products/kepserverex/opc-interoperability/ (Accessed 10 Feb 2023)
12. Chehri, A., Mouftah, H., Fortier, P., Aniss, H.: Experimental testing of IEEE801.15.4, ZigBee sensor networks in confined area. In: 8th Annual Communication Networks and Services Research Conference, Montreal, QC, Canada, vol. 2010, pp. 244–247 (2010). https://doi.org/ 10.1109/CNSR.2010.62

Intelligent Transport Systems

Communication Trends, Research Challenges in Autonomous Driving and Different Paradigms of Object Detection

Teena Sharma[1], Abdellah Chehri[1(✉)], and Paul Fortier[2]

[1] Department of Mathematics and Computer Science, Royal Military College (RMC), Kingston, Canada
teena.sharma@rmc-cmr.ca, chehri@rmc.ca
[2] Department of ECE, Laval University, Quebec City, QC G1V 0A6, Canada
fortier@gel.ulaval.ca

Abstract. Autonomous vehicle (AV) technology has the potential to provide a secure, robust and easy mode of transportation for the general public. A connected autonomous vehicle (CAV) is an AV that has vehicle communication capability, which improves the AV's situational awareness and allows it to collaborate with other AVs. As a result, CAV technology will improve the capabilities and robustness of AV, making it a promising potential transportation solution in the future 5G. This paper introduces a representative architecture of CAVs and surveys the latest research trends for 5G. It reviews the state-of-the-art and state-of-the-practice of the latest literature on autonomous vehicles (AV) technologies in 5G and it studies technologies trends and key technologies for Autonomous Driving. The main issues and unresolved problems are critically discussed based on the reviews to determine potential research directions. We have also presented different paradigms of Object Detection in the field of Autonomous driving.

Keywords: Autonomous Driving · Smart City · Autonomous Vehicle · Machine Learning · 5G · Intelligent Transportation Systems

1 Introduction

A connected car is a vehicle that allows the exchange of data and information [1]. The connected car will accompany our daily life on several points according to the degree of connectivity. They are distributed in three different topics:

- *Intra-connectivity:* This is the connectivity that remains in the vehicle, Bluetooth, real-time driving analysis. Example: consumption monitoring or eco-driving to display the recommended gear change at the right time.
- *The extra-connectivity:* It designates the connections made with the outside, by the GPS coordinates, etc. and gives the possibility to an organization to receive information on the car, the number of kilometers traveled, etc. This can be useful if you subscribe to insurance per kilometer, for example, to track the number of kilometers traveled more easily.

© The Author(s), under exclusive license to Springer Nature Singapore Pte Ltd. 2023
A. Zimmermann et al. (Eds.): KES-HCIS 2023, SIST 359, pp. 57–66, 2023.
https://doi.org/10.1007/978-981-99-3424-9_6

- *Inter-connectivity:* The highest degree of connectivity but be careful not to confuse a connected car with an autonomous car. Here, data is exchanged in both directions to benefit all drivers of a connected vehicle. Real-time traffic information allows you to avoid traffic jams and traffic jams by offering you an alternative route to avoid them. They will enable the user to be warned of all dangers on the road reported by other drivers, or a GPS beacon hidden in the car allows you to locate it in case of theft and even deactivate the engine for certain vehicles. With evolution and rapid growth in 5G (5th Generation mobile networks) technologies, very low latency and large spectral efficiency is being expected while efficiently managing high amount of traffic without interference. A 5G Technology is a user-centric network model as well as a modern access technology. Its mission is to provide a single network and multiple services to customers using all available and imagined technologies rather than modifying current communication architectures (e.g., LTE) [2].

The real-time transmission of sensor and location data, the uploading and downloading of vast amounts of data in the cloud, and even the transmission of entertainment video and advertising all necessitate higher network bandwidth and lower network latency in autonomous driving. Furthermore, autonomous vehicles can drive at extremely high speeds and in close proximity to one another. As a result, autonomous vehicles' communication criteria are more stringent than conventional vehicles, whether in terms of latency, reliability, scalability, or versatility. 5G technology currently uses the existing LTE frequency spectrum and millimeter-wave bands (24–86 GHz), as well as NOMA technology in order to enhance the spectral performance.

It can accommodate the 256/1024QAM amplitude modulation format and maintain an end-to-end network latency of less than 10 ms [3, 4]. The new 5G technology has sparked a lot of interest. In [5], authors suggested a software-defined cloudlet for scheduling management and transmission in a 5G vehicular network connectivity scheme.

In [6] authors present a 5G SDN-based vehicular network that includes a fog unit to cover the vehicle flexibly and prevent frequent vehicle handover between roadside units and vehicles. Its transmission delay and throughput are investigated, and the findings show that the planned scheme has the shortest transmission delay. In [7] researchers suggested that 5G's unique signal characteristics are ideal for vehicle positioning and analyze the cellular and 5G mmWave positioning implementation processes. Eiza et al. [8] also suggest a new framework model for 5G-enabled vehicle networks that is conducive to dependable, stable, and privacy-conscious real-time video reporting services. A short summary of these proposed schemes is presented in Table 1.

The arrival of connected and autonomous vehicles, as well as 5G, will revolutionize our daily lives. The autonomous car will, for example, be able to communicate with red lights in order to regulate its speed so as not to have to stop at a red light and thus facilitate traffic flow. Other examples of changes are the automation of the gate and garage door opening when the car arrives, the communication between the vehicles to adapt their speed [9].

This paper surveys the latest literature on CAV technologies in 5G and studies sensor fusion algorithms, perception, planning, and control functions. The main issues and unresolved problems are discussed based on the reviews to determine potential research

Table 1. Comparative analysis of various Technologies in autonomous driving

Technology	Bit rate	Standard	Spectrum	Modulation	Delays	Range	Modes
VLC	500 Mbps	IEEE 802.15.7	430–790 THz	OOK, VPPM	Very low	100 m	N/A
5G	10 Gbps	NA	24–86 GHz, 600 MHz-6 GHz	NOMA, Massive MIMO	1ms	2 km	NA
LTE-V	1 Gbps	LTE-V	N/A	SC-FDMA, MIMO, OFDMA	50 ms	2 km	LTE-V-Cell LTE-V-Direct
WiMax	128 Mbps	IEEE 802.16	2.5 GHz	OFDMA, MIMO	10 ms	50 km	Mesh, PMP
UWB	10 Mbps	IEEE 802.15.3a	3.1–10.6 GHz	MB-OFDM	N/A	10 m	N/A
WiFi	600 Mbps	IEEE 802.11n	5.150–5.850 GHz, 2.4–2.483 GHz	OFDM, MIMO	Seconds	100 m	WDS, Mesh, access point, STA, Monitor
	1 Gbps	IEEE 802.11ac	5 GHz	OFDM, MIMO	Seconds	100 m	WDS, Mesh, access point, STA, Monitor
Bluetooth	1–24 Mbps	IEEE 802.15.1	2.4–2.485 GHz	8-DPSK, GFSK, DQPSK	3–10 s	100 m	Hold, Active, Sniff, Park
ZigBee	20 Kbps, 40 Kbps, 250 Kbps	IEEE 802.15.4	868 MHz, 915 MHz, 2.4 GHz	O-QPSK, BPSK	30 ms	100 m	N/A
DSRC	54 Mbps	IEEE 802.11p	5.850–5.925 GHz	OFDM	100 ms	1 km	Active, Passive

directions. The rest of the paper is organized as follows: Sect. 2 surveys the key computing technologies, AI and Deep Learning Applications in autonomous driving, Sect. 3 presents different paradigms of object detection in autonomous environment. Section 4 illustrates the challenges and open issues in future autonomous vehicle technology. Finally, Sect. 5 concludes the review paper with future recommendations.

2 Connected Autonomous Vehicle Architecture

In a nutshell, CAV entails three key tasks, as seen in Fig. 1, namely data acquisition (awareness), Data processing (preparation), and actuation (power). The perception layer collects data from various sources to track the atmosphere. The perception layer measures the global and local position of the ego-vehicle and creates a map of the environment using data directly from sensors or sensor fusion techniques. Based on remote map data of road and traffic details, the planning layer determines the best global route from its current location in the world to the to the requested destination. The planning layer then computes a local optimal trajectory using online decision making and trajectory planning based on real-time vehicle states and the current environment (e.g., lane markers, traffic, pedestrians, road signs, etc.) given by the perception layer.

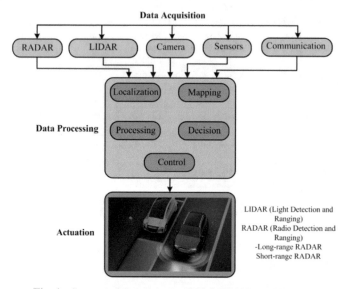

Fig. 1. Connected Autonomous Vehicle (CAV) Architecture

The planning layer then computes a local optimal trajectory using online decision making and trajectory planning based on real-time vehicle states and the current environment (e.g., lane markers, traffic, pedestrians, road signs, etc.) given by the perception layer. Finally, the control layer determines the required order to control the actuators (e.g., steering wheel, gas pedal, brake pedal, etc.) in the vehicle to follow the desired trajectory [10]. The perception layer can share its perception data with other road users, and the planning layer can conduct cooperative driving with other road users, thanks to vehicle connectivity. The decision-making algorithms are used in the planning phase as part of a hierarchical control process, with decisions made at a higher-level affecting servo control at a lower level to control or actuate the vehicle. Among the methods to improve safety, we distinguish external methods linked to the infrastructure, roads, and signaling quality. But also, the internal vehicle method, with systems to prevent wheel spin, lane-keeping assistance, and collision avoidance. Lately, there has been an increase in the number of accidents, which can have multiple causes [11–13]. Many of these accidents were caused by a collision between two or more vehicles that were preventable. The autonomous car is equipped with several sensors which analyze all the data emanating from the vehicle and its road environment. To drive safely, the autonomous vehicle has a whole arsenal of sensors (Fig. 1):

- *Lidars:* these laser scanners are designed to detect the surroundings of the car and prevent possible dangers thanks to a laser light beam returned to its emitter. For autonomous cars, LIDAR tracking is used. The main characteristic of LIDAR includes 360-degree visualization and object tracking with a relatively long range. Therefore, a LIDAR device can be mounted atop the car to get a full view of the surrounding environment.

- *Ultrasonic sonars:* *t*hese sensors validate the information provided by laser scanners and make it possible to detect objects close to the vehicle.
- *Cameras:* Those that equip self-driving cars have a 360-degree field of view. They allow the vehicle to respect the highway code by detecting red lights and all other traffic signs.
- *Radar:* This technology has proven to be more efficient in object tracking than cameras, making it a more practical option for vehicles. EM waves determine the relative speed and relative position of the detected obstacles [14], The Doppler effect, also known as Doppler shift, refers to the variations or shifts in wave frequency arising from relative motion between a wave source and its targets. For intensive object detection such as collision-resistance while parking, collision avoidance, and bumper protection, LIDAR does not work efficiently. Instead, optimized radars are installed at the front, rear, and sides of the car for the aforementioned tasks. Key Technologies in Autonomous Driving.

2.1 Computing Technologies

Computing technologies are needed to deal with tasks that are difficult for autonomous vehicles to complete themselves due to the restricted storage and computing capacities of their on-board terminals. The cloud's processing resources can be put to work on computation-intensive tasks for self-driving cars. However, this approach necessitates a large amount of bandwidth to transfer raw data, which will result in increased latency between autonomous vehicles and the cloud. As a result, the cloud is well suited to long-term and non-real-time autonomous driving tasks, such as large-scale model training, the generation and updating of high-precision maps, and vast data storage and simulation.

Furthermore, autonomous vehicles need to make quick decisions about their control, conduct route planning, process perceptual data in real time, and take appropriate driving action, while edge computing, as one of the 5G innovations and the extension of cloud computing to edge networks, will provide adjacent services for a variety of time-sensitive processing needs. It can handle real-time tasks like perception analysis and sensor data fusion in autonomous vehicles [15–17].

2.2 AI and Deep Learning Application in Autonomous Vehicle

Autonomous driving systems can reliably distinguish routes, pedestrians, obstacles, and other factors in order to make the best decision possible. However, whether in bodywork or networking and communications, deep learning technology will have a brighter future in autonomous driving. Deep learning technology extracts features from large-scale and high-dimensional datasets using a cascade of several layers of nonlinear processing units [18]. It has important applications in the conventional computer vision field of autonomous driving, as the abundance of data from sensors such as Manual modelling of LIDAR, Radar, and Camera is difficult.

3   Object Detection in Autonomous Environment

Object detection plays a crucial role to regulate traffic situations in autonomous environments. It can extract useful and reliable traffic information, which may be used for traffic image analysis and the regulation of traffic flow. This information comprises the number of vehicles, their trajectories, where they are being tracked, the flow of vehicles, the classification of vehicles, the traffic density, the vehicle velocity, the changes in traffic lanes, and the identification of license plates [16]. In addition, the information can be used to assist in the detection of other road assets, including pedestrians, different types of vehicles, people, traffic lights, earthworks, drainage, safety barriers, signs, lines, and the soft estate, which includes grassland, trees, and shrubs, by making use of various object detectors.

Object detection in normal or autonomous environment may be affected by bad weather conditions such as hue or if it's too snowy or rain [17]. Due to changing environmental or weather conditions such as snowstorms, hue, rain and sunny it become unpredictable for the driverless cars and even for driver to predict traffic conditions. In addition, clear object recognition is hard and therefore it leads to wrong judgement of vehicles or other objects on the road. In such cases, various prediction-based previously trained models and algorithms are used to provide proper judgement. Moreover, information about road damage by excessive rain or snow and ongoing road construction work can be determined by using efficient object detection techniques. In all these situations, pre alerts must be sent to drivers or autonomous vehicles at the appropriate time so that driver can choose a different path and save time as well as can avoid any hazards. Different object detection techniques enable for the determination of varying traffic predictions, and based on those predictions, alerts or warning messages can be sent to drivers or autonomous cars.

There have been many research and surveys that have provided numerous object identification strategies that can be used in vehicular environments. Out of these techniques, the three most prevalent detection methodologies are manual, semi-automatic, or completely automated [18]. Manual and semi-automated surveys are the two ways that have traditionally been used in the process of collecting data on the many objects that may be found on roadsides. In the manual method, a visual inspection of the things that are present on the streets and roads is accomplished either by strolling along the streets and roads or by driving along the streets and roads in a vehicle that moves at a slow speed. This type of examination is plagued by the inspectors' propensity for making subjective assessments. It requires a significant human intervention that is proven to be time-consuming, given the extensive length of road networks and number of objects. Moreover, inspectors must often be physically present in the travel lane, exposing themselves to potentially hazardous conditions.

In semi-automated object detection procedures [19], the objects on the roads/streets are collected automatically from a fast-moving vehicle, but the collected data is processed in workstations at the office. This approach improves safety but still is based on approach, which is very time-consuming. Fully automated object detection techniques often employ vehicles equipped with high-resolution digital cameras and sensors [20].

Many CNN designs have been created to provide the greatest accuracy with increased processing speed. Most popular and widely used CNN techniques are: R-CNN (Region-based Convolutional Neural Networks [21], Fast-RCNN [22] and Faster-RCNN [23].

However, the computational load was still too large for processing images on devices with limited computation, power, and space.

Real-time road lanes detection and tracking Automatic tracking of the road is probably one of the approaches that have generated the most work in recent decades. Its implementation is based on two complementary modules ensuring the following tasks:

- The detection of the lines separating the different lanes of the roadway.
- The detection and location of obstacles possibly found on the current trajectory of the vehicle.

Indeed, the detection and location of obstacles is an integral part of the task of automatic tracking of the road. As a result, these two research themes in the field of intelligent transport have forged a common destiny. In what follows, we discuss in more detail the detection of markings on the ground and the detection of obstacles. In automated vehicles, the lanes detection and tracking are divided into two stages of the "perception" and "action". Some examples of the strategies that can be adopted to solve the problem of lateral control are presented in [24]. Be that as it may, the position of the markings on the road being useful to other subsystems of comparable devices, the location of these is in any case generally executed. Currently, the CAVs are at stage two, which means that some driver assistance systems can be automated, such as cruise control and lane-centering.

4   Challenges in Future Autonomous Vehicle Technology

Autonomous vehicles have more sophisticated functions than conventional vehicles, such as dynamic route planning, real-time driving activity customization, and adaptive scene mode switching. Since autonomous vehicles have minimal onboard storage and processing power, they will be more reliant on the cloud or edge. As a result, data storage, authentication, network transmission, and virtualization are all significant challenges in autonomous vehicle technology in terms of security [25].

Figure 2 shows presents brief flowchart of research challenges and open issues as demonstrated:

1. *Data sharing:* In conventional vehicle networks, multimedia infotainment is heavily reliant on service providers. In order to gather more real data and make more decisions, future autonomous vehicle networks would need a large number of autonomous vehicles to engage in data sharing. As a result, it is dependent on the number, better predictions and popularity of autonomous vehicles to make correct decisions. If autonomous vehicles are not widely used and the number of connected autonomous vehicles is small, the rate of information sharing will be poor, posing a significant challenge to autonomous driving communications [26].

2. *Information Transmission Priority:* As the number of potential autonomous vehicles grows and the gap between them shrinks, their distribution will become increasingly dense, and the number of messages to be transmitted will increase. Since channel bandwidth is restricted and the autonomous vehicle's driving process is primarily determined by the channel, he priority of transmitted information in an autonomous driving system must be carefully developed to ensure that Self-driving cars can get

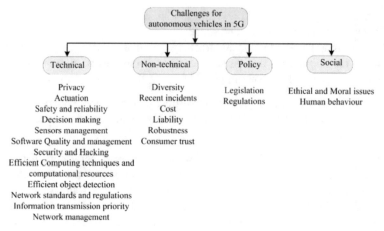

Fig. 2. Various challenges and open issues in autonomous vehicle technology

timely access to data that is valuable to them. Additionally, too much focus on the transmission of such messages will greatly weaken the transmission of environmental information, which is also the foundation of autonomous driving. To improve this, the priority of various information transmissions needs to be weighed very carefully.

3. *Security issues:* Security concerns are key at all times and in all places, especially when it comes to autonomous driving, which is directly linked to human life. The underlying structure of autonomous vehicles controls all aspects of their operation without the need for human interference. Hackers can use wireless network attack methods like brute force cracking and packet capture to spread viruses and Trojans across the network, to attack autonomous vehicles [27]. As a result, not only will people's personal details be compromised, but autonomous vehicles could also be operated remotely, paralysing the entire autonomous vehicle network, resulting in a series of tragedies. As a result, it's essential to isolate the autonomous vehicle's bottom layer and accelerate the creation of a driverless firewall.

4. *Network management:* As the number of mobile data and networking devices grows across the world, network capacity shared among people, autonomous vehicles, and other entities becomes insufficient, resulting in network resource rivalry. Apart from this, more complex infrastructures and faster autonomous vehicles leads to regular handover issues, resulting in constant interaction between vehicles or between vehicles and infrastructures, resulting in serious interference [28].

5. *Growth in network Standards and regulations:* Autonomous driving possesses significant cross-industry and cross-field characteristics. However, it is still in the early stages of growth, its business model is uncertain, and there are still gaps between different fields. Manufacturers of commercial vehicles, Internet providers, telecommunications companies, states, and other public entities have yet to achieve effective integration [29].

6. *Developments in computing technology:* The advancement of autonomous driving necessitates the integration of sensor, transmission, and data processing technologies, as well as the convergence of multi-source data, necessitating virtualization,

mass storage, and computational capabilities. Since, the cloud computing data centre is situated in the core network to support autonomous vehicles from various geographical locations. However, since it is located so far away from the end user, it suffers from high latency, network congestion, and poor reliability [30].

5 Conclusion and Perspective

With the rapid advancement in technology and scientific and engineering studies, the realization of CAV is not far off from being a reality in our everyday lives. The main advantages of self-driving vehicles include new market prospects, improved protection for both passengers and outsiders (pedestrians and other vehicles), ease of use and comfort, improved traffic conditions, and a desire to drive a customer-centered experience. Despite the numerous advantages, it has many research issues and challenges, such as regulations, security threats, safety and reliability, authentication, encryption, standards, network management, and computing techniques. These challenges must be resolved before commercial autonomous vehicles can be completely deployed on the roads. In this paper, we have reviewed existing state-of-the-art technologies, design, and implementation problems of CAVs.

References

1. Hussain R., Zeadally S.: Autonomous Cars: Research Results, Issues, and Future Challenges. IEEE Commun. Surv. Tutor **21**(2), 1275–1313, Second quarter (2019)
2. Almirall, E., et al: A. Smart cities at the crossroads: New tensions in the city transformation. Calif. Manag. Rev. **59**, 141–152 (2017)
3. Domingo, A., Bellalta, B., Palacin, M., Oliver, M., Almirall, E.: Public open sensor data: Revolutionizing smart cities. IEEE Technol. Soc. Mag. **32**, 50–56 (2013)
4. Ji, Z., Ganchev, I., O'Droma, M., Zhao, L., Zhang, X.: A cloud-based car parking middleware for IoT-based smart cities: design and implementation. Sensors (2014)
5. Chehri, A., Mouftah, H.T.: Autonomous vehicles in the sustainable cities, the beginning of a green adventure. Sustainable Cities Soc. (SCS) **51**, Article 101751 (2019)
6. Chehri, A., Mouftah, H.T.: Localization for Vehicular Ad Hoc network and autonomous vehicles, are we done yet? Connected and Autonomous Vehicles in Smart Cities, CRC Press, Taylor & Francis Group (2020)
7. Song, Y., Liao, C.: Analysis and review of state-of-the-art automatic parking assist system. In: IEEE International Conference on Vehicular Electronics and Safety, pp. 1–6 (2016)
8. Chehri, A., Quadar, N., Saadane, R.: Communication and localization techniques in vanet network for intelligent traffic system in smart cities: a review. In: Qu, X., Zhen, Lu., Howlett, R.J., Jain, L.C. (eds.) Smart Transportation Systems 2020. SIST, vol. 185, pp. 167–177. Springer, Singapore (2020). https://doi.org/10.1007/978-981-15-5270-0_15
9. Chehri, A., Sharma, T., Debaque, B., Duclos, N., Fortier, P.: Transport systems for smarter cities, a practical case applied to traffic management in the city of montreal. In: International Conference on Sustainability in Energy and Buildings, SEB-21, KES Virtual Conference Centre, pp. 15–17 (September 2021)
10. Chehri, H., Chehri, A., Saadane, R.: Traffic signs detection and recognition system in snowy environment using deep learning. In: Ben Ahmed, M., Rakıp Karaş, İ, Santos, D., Sergeyeva, O., Boudhir, A.A. (eds.) SCA 2020. LNNS, vol. 183, pp. 503–513. Springer, Cham (2021). https://doi.org/10.1007/978-3-030-66840-2_38

11. Zang, S., et al.: The impact of adverse weather conditions on autonomous vehicles: how rain, snow, fog, and hail affect the performance of a self-driving car. IEEE Veh. Technol. Mag. **14**(2), 103–111 (2019)
12. Liu, L.C., Xie, D., Wang, S., Zhang, Z., CCN-based cooperative caching in VANET. In: 2015 International Conference on Connected Vehicles and Expo, pp. 198–203 (2015)
13. Guo, H., Liu, J.: Collaborative computation offloading for multiaccess edge computing over fiber–wireless networks. IEEE Trans. Veh. Technol. **67**, 4514–4526 (2018)
14. Guo, H., Liu, J., Zhang, J.: Computation offloading for multi-access mobile edge computing in ultra-dense networks. IEEE Commun. Mag. **56**, 14–19 (2018)
15. Sun, W., Liu, J., Yue, Y., Zhang, H., Double auction-based resource allocation for mobile edge computing in industrial internet of things. IEEE Trans. Ind. Informatics **14**, 4692–4701
16. Wang, J., Liu, J., Kato, N.: Networking and communications in autonomous driving: A survey. IEEE Commun. Surv. Tutorials **21**, 1243–1274 (2018)
17. Pendleton, S.D., et al.: Perception, planning, control, and coordination for autonomous vehicles. Machines (2017)
18. Schmidhuber, J.: Deep learning in neural networks: an overview. Neural Netw. **61**(2015), 85–1172 (2014)
19. Sharma, T., Debaque, B., Duclos, N., Chehri, A., Kinder, B., Fortier, P.: Deep learning-based object detection and scene perception under bad weather conditions. Electronics **11**(4), 563 (2022)
20. Redmon, J., Farhadi, A.: YOLOv3: An Incremental Improvement. Tech. rep. (2018). arXiv: 1804.02767v1
21. Zhang, S., et al.: Single-shot refinement neural network for object detection" In: (2017). DOI: https://doi.org/10.1109/CVPR.2018.00442. arXiv: 1711.06897
22. Chehri, A., Quadar, N., Saadane, R.: Survey on localization methods for autonomous vehicles in smart cities. In: Proceedings of the 4th International Conference on Smart City Applications (SCA 2019), Article 113, pp. 1–6. Association for Computing Machinery, New York (2019). https://doi.org/10.1145/3368756.3369101
23. Sheehan, B., Murphy, F., Mullins, M., Ryan, C.: Connected and autonomous vehicles: A cyber-risk classification framework. Transp. Res. Part A Policy Pract. **124**, 523–536 (2019)
24. Maple, C., Bradbury, M., Le, A.T., Ghirardello, K.: A connected and autonomous vehicle reference architecture for attack surface analysis. Appl. Sci. **9**, 5101 (2019)
25. Cunneen, M., Mullins, M., Murphy, F., Shannon, D., Furxhi, I., Ryan, C.: Autonomous vehicles and avoiding the trolley (Dilemma): Vehicle perception classification and the challenges of framing decision ethics. Cybern. Syst. **51**(1), 59–80 (2020)
26. Cunneen, M., et al.: Autonomous vehicles and embedded artificial intelligence: The challenges of framing machine driving decisions. Appl. Artif. Intell., 33(8), (2019)
27. Ryan, C., Murphy, F., Mullins, M.: Semiautonomous vehicle risk analysis: A Telematics-based anomaly detection approach. Risk Anal. **39**(5), 1125–1140 (2018)
28. Ryan, C., Murphy, F., Mullins, M.: end-to-end autonomous driving risk analysis: a behavioural anomaly detection approach. IEEE Trans. Intell. Transp. Syst. **22**(3), 1650–1662 (2021)
29. Eskandarian, A., et al. Research advances and challenges of autonomous and connected ground vehicles. IEEE Trans. Intell. Trans. Syst. **22**(2), 683–711
30. Chehri, A., Chehri, H., Hakim, N., Saadane, R.: Realistic 5.9 GHz dsrc vehicle-to-vehicle wireless communication protocols for cooperative collision warning in underground mining. In: Qu, X., Lu., Zhen, Howlett, R.J., Jain, L.C. (eds.) Smart Transportation Systems 2020. SIST, vol. 185, pp. 133–141. Springer, Singapore (2020). https://doi.org/10.1007/978-981-15-5270-0_12

Investigation in Automotive Technologies Transitions

Milan Todorovic$^{(\boxtimes)}$ ⓘ, Abdulaziz Aldakkhelallahⓘ, and Milan Simicⓘ

RMIT University, Melbourne 3000, Australia
milan.todorovic@yahoo.com

Abstract. Research on autonomous vehicles (AV) and electric vehicles (EV) introduction management is reviewed bibliometrically and presented in this paper. In the investigation, we discuss technology changes' primary traits, evolution, and a number of transitional problems that appear when going from conventional vehicles to AVs and EVs. We have identified possible trends and following that, directions for further studies. Understanding more general developments in the automotive industry, such as those related to sustainability, environmental protection, traffic safety, market factors, public policies, new business models and other management aspects, was necessary to perform the conducted analysis. This review identified numerous research gaps. The Scopus and WoS search for relevant articles generated 4693 articles analyzed using the Vosviewer visualization software.

Keywords: Bibliometric Review · Autonomous Vehicles · Electric Vehicles · Transition · Scopus

1 Introduction

This literature review aims to present research on the transition to autonomous and electric vehicles, particularly on work published in the previous six years. Review differs from other literature studies in that it aims to present only a selection of the works from the beginning of the research of specific topics and to choose only the most influential works that can inspire further research. In addition, the goal was to identify the subjects that attract the most interest from researchers, who influence industry stakeholders and regulators.

An increased interest in topics relating to autonomous and electric vehicles can be determined by examining the number of published articles. The number of publications can attest to the rise in academic interest in studying related subjects.

This paper aims to address the need for electric vehicles in determining new transportation paradigms given the complexity of the transition to autonomous and electric vehicles and the growing demand for a new concept that emphasizes sustainability. Additionally, our goal was to conduct a literature bibliometric analysis of the effects of autonomous and electrical vehicles on various social issues.

We first present the study methodology and discuss the data collection and analysis methods after giving a brief overview of autonomous and electric vehicles. Second, the

© The Author(s), under exclusive license to Springer Nature Singapore Pte Ltd. 2023
A. Zimmermann et al. (Eds.): KES-HCIS 2023, SIST 359, pp. 67–77, 2023.
https://doi.org/10.1007/978-981-99-3424-9_7

findings comprise results on the bibliographic links and descriptive evidence regarding the research sample. Thirdly, research gaps were identified by comparing the findings with a larger body of knowledge. The final section concludes by outlining significant implications and suggesting strategies for future research.

2 Autonomous and Electric Vehicles Overview

Electric and autonomous vehicles are now frequently seen on our roads. However, to make the transition from old to new technology, there are still many issues that need to be resolved. In addition, we should also be aware that old and new technologies will coexist in a few decades. Because of this, the transition process is even more difficult.

Future technologies are predicted to have a wide range of effects on the labor market, the health and welfare systems, urban planning, transportation, and the automotive sector. Self-driving vehicles are one component of the connected, autonomous, and shared, mobility vision for the future of transportation, including other technologies like vehicle electrification. They are subsystems of future Intelligent Transport Systems (ITS).

Automobiles with integrated vehicle automation are self-driving cars, also referred to as autonomous, driverless, or mobile robots [1–3]. With little human intervention, ground vehicles can sense their surroundings and move safely [4]. Self-driving cars use a variety of sensors, including infrared cameras, radar, lidar, sonar, GPS, odometry, and inertial measurement units, to perceive their surroundings [2]. In addition, advanced control systems analyze sensory data to determine the best routes, obstacles to avoid, and pertinent signs [5–8]. The fact that GPS could be unreliable in busy inner-city settings is a technological barrier to the widespread adoption of automated vehicles.

Numerous surveys were conducted where consumers expressed their interest in purchasing or investing in new technology, their perception of safety, comfort, cybersecurity, privacy, and the implications of different automation levels. Adoption levels are investigated in the studies with participants from Australia and Saudi Arabia [9–11].

A vehicle propelled by one or more electric motors using the stored energy in batteries is referred to as an electric vehicle. Electric-powered motors are quieter, produce no exhaust emissions, and reduce overall emissions compared to cars powered by internal combustion engines (ICE). Due to the decline in fuel and maintenance costs, as of 2020, the total cost of ownership of the most recent electric-powered vehicles will be lower than that of comparable ICE vehicles. Governmental incentives for plug-in electric vehicles are implemented in many countries, including tax credits, subsidies, and other non-economic incentives. To reduce air pollution and slow climate change, many nations have passed legislations mandating the gradual phase-out of the sale of fossil fuel-powered vehicles.

3 Research Methodology

Due to its greater material coverage than other databases, Elsevier's Scopus database was selected as the primary source to look for publications on the research topic. In addition, it is descriptive and employs a quantitative method to pinpoint crucial traits, the development of AV/EV domains, and trends for upcoming studies.

Study objectives were to a) evaluate the number of scholarly publications pertinent to the transition generally and b) conduct scientometric and literature analyses to characterize the studies that have received the most citations in the area. Thus, the following are some benefits of the bibliometric analysis findings:

a. giving the research problem clarity and focus
b. measuring and comprehending a particular subject's study area
c. giving a clear picture of the historical development of the field
d. improving the research process and the depth of knowledge
e. presenting a technological and thematic analysis and
f. putting the results into context and laying the groundwork for additional study.

The current study suggests four stages, described in the following steps:

Stage 1: Determining the Scope of Analysis and Article Selection

The first two stages are choosing the articles for analysis and defining the analysis's scope. EVs and AVs are discussed, but the transition is the main topic. In the PRISMA 2020 statement we have an updated guideline for reporting systematic reviews [12]. The procedure for selecting relevant literature is shown in Fig. 1. We will look for keywords in titles, abstracts, author keywords, publisher-defined keywords, and keywords used by publishers to categorize works. Only academic, peer-reviewed articles and reviews, not books or book chapters, are included in this summary.

Stage 2: A Descriptive Analysis of the Work

The following analysis was done: (1) the number of published articles per search term; (2) the Annual number of papers; (3) the Most published authors; (4) Most public sources; and (5) the Countries analyzed.

Stage 3: A Thorough Examination of the Papers.

Based on a literature search using the Vosviewer bibliometric visualization software the following bibliographic literature review parameters were considered:

- Analysis of the primary Scopus categories in which the articles were published.
- Analysis of the most significant keywords and terms.
- Dual-map overlay, institutional productivity, author productivity, corresponding author analysis, top citation analysis, and co-citation analysis were.
- The total number of papers that authors produced, the analysis of citations and the co-occurrence of terms.

Stage 4: Interpretation and Discussion of the Findings

The interpretation and discussion of the results were aided by the bibliometric visualization tool Vosviewer. The findings highlight the major research trajectories and knowledge gaps in various fields. Results are shown graphically and in tabular form.

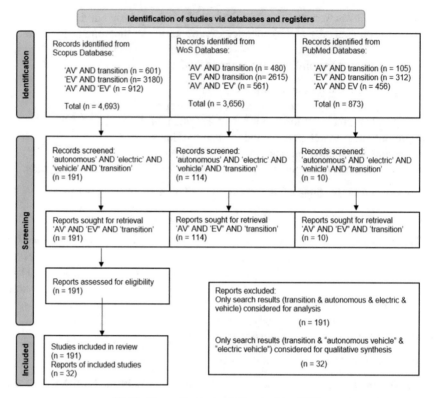

Fig.1. Research approach (inspired by [12])

4 Results

Regarding general survey data, we found that the papers researching the simultaneous transition to autonomous and electric vehicles started to be presented in 2017.

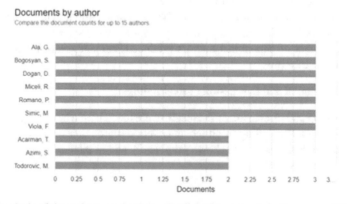

Fig. 2. Analysis of the authors productivity: Published documents in the group of 191 articles

In the last three years their numbers increased from 3 (2019, 2020) to 8 (2021) and 6 (2022). Figure 2. Shows an analysis of the author's productivity. By analyzing the

Table 1. Analysis of the most cited papers

Paper	Authors	Citations	Reference
Functional, symbolic and societal frames for automobility: Implications for sustainability transitions	Sovacool, B.K., Axsen, J.	54	[13]
Analysis of consumer attitudes towards autonomous, connected, and electric vehicles: A survey in China	Wu, J., Liao, H., Wang, J.W.	37	[14]
Comparing technology acceptance for autonomous vehicles, battery electric vehicles, and car sharing - A study across Europe, China, and North America	Muller, J.M	30	[15]
The electrification accelerator: Understanding the implications of autonomous vehicles for electric utilities	Weiss, J., Hledik, R., Lueken, R., Lee, T., Gorman, W.	25	[16]
Combining analytics and simulation methods to assess the impact of shared, autonomous electric vehicles on sustainable urban mobility	Dlugosch, O., Brandt, T., Neumann, D.	21	[17]
Comprehensive analysis method of determining global long-term GHG mitigation potential of passenger battery electric vehicles	How, F., Chen, X., Chen, X., Yang, F., Ma, Z., Zhang, S., Liu, C., Zhao, Y., Guo, F.	18	[18]
Managing Transition to Electrical and Autonomous Vehicles	Todorovic, M., Simic, M., Kumar, A.	16	[19]
Comparing the effects of vehicle automation, policymaking and changed user preferences on the uptake of electric cars and emissions from transport	Mazur, C., Offer, G., Contestabile, M., Brandon, N.B.	16	[20]
Machine-learning based approaches for self-tuning trajectory tracking controllers under terrain changes in repetitive tasks	Prado, Á.J., Michałek, M.M., Cheein, F.A.	13	[21]

(*continued*)

Table 1. (*continued*)

Paper	Authors	Citations	Reference
A novel approach for plug-in electric vehicle planning and electricity load management in presence of a clean disruptive technology	Hajebrahimi, A., Kamwa, I., Huneault, M.	12	[22]

broadened group of 191 articles, the authors with the most articles are Ala G, Miceli R, Romano P, Viola F, and Simic M. Table 1 presents analyses of the leading publications which have the highest number of citations.

The work of Sovacool, B.K. et al., *Functional, symbolic and societal frames for automobility: Implications for sustainability transitions*, [13] has been cited 54 times, making it the most cited paper of the transition to autonomous and electric vehicles research. This paper's authors develop a conceptual framework that explores automobility through categorizing frames or shared cultural meanings.

The second most cited work is by Wu, J. et all., *Analysis of consumer attitudes towards autonomous, connected, and electric vehicles: A survey in China*, [14] with a total of 37 citations. The authors aim to understand consumer attitudes towards autonomous, connected, and electric vehicles, using data collected through a survey in China. They found the potential for environmental-friendly transport, increased accessibility of travel for non-drivers, and reduced driving fatigue as the most attractive aspects. The biggest concern is related to safety, legal liability and charging issues.

The third most cited publication is Muller, J.M. et all., *Comparing technology acceptance for autonomous vehicles, battery electric vehicles, and car sharing* - A study across Europe, China, and North America, with 30 citations. In this paper, the authors surveyed more than 1000 participants to establish their acceptance of AV/EV technology and car-sharing and the significance of these findings to the industry, society and policymakers [15].

The work by Todorovic et al. *Managing Transition to Autonomous and Electric Vehicles Using Fuzzy Logic*, describes the decisions on the optimal pathways which depend on the big data, incomplete and inconsistent knowledge, and expectations [23]. Presented is the application of fuzzy logic in the decision-making process of the transition to AV. It considers the crisp and fuzzy information, outcomes, and actions and compares additional information values. The Bayesian framework, transformed into a toolbox, can be used in any other fuzzy domain.

A bibliometric literature analysis is a standard and rigorous method for investigating and analyzing large amounts of scientific data. Bibliometric methods employ a quantitative approach for describing, evaluating, and monitoring published research. Scientific publications are used as a data source to provide a better understanding of how research is created, organized, and linked.

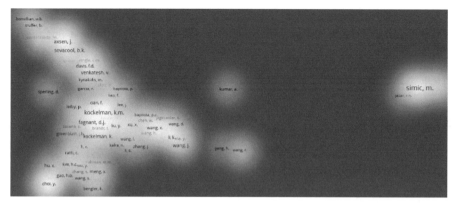

Fig. 3. Analysis of the authors' co-citations (density visualization minimum 3 co-citations)

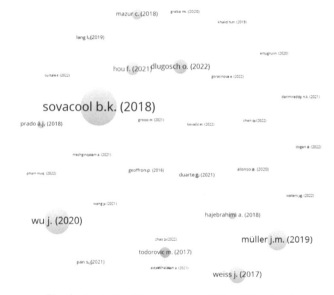

Fig. 4. Analysis of the papers by citation (32 authors)

This analysis was undertaken to understand research trends and to assess citations as a measure of impact. The citation of articles can be observed by visualization. In this analysis, we will use Vosviewer software.

Figure 3. Presents the co-citation density visualization analysis. It shows authors and the impact of their research on other authors and their studies. The network visualization uses circles to indicate authors and their works. Clusters define the similarity of the research areas. The analysis presented in Fig. 4 shows the citations of the papers, which have at least one citation. The larger the size of the circle, the greater the citation and the greater the influence of this paper and its authors.

Another bibliographic analysis of the most cited works shows the research content in that area. It is a co-occurrence analysis. This data shows us the areas where the research was conducted. Moreover, we can follow Fig. 5 to see when certain concepts were first mentioned and when they started to be studied. Those common co-occurrence terms are extended lists of authors' and publishers' used keywords and terms.

5   Discussion

Autonomous vehicles have reached the level that further testing, and verification of technological readiness must be conducted on the road. Moreover, over time it has been realized that infrastructure planning in cities and on the open road will have to be done studiously and include different stakeholders.

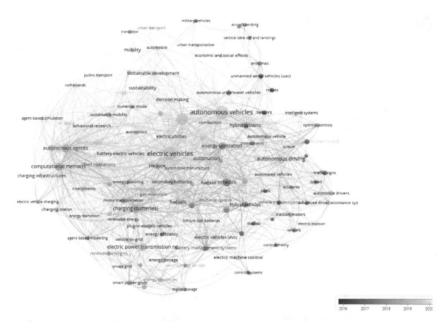

Fig. 5. Analysis of the papers by terms co-occurrence (overlay, 191 articles, 4 terms)

The main reason for the introduction of electric vehicles was the fear that fossil fuels would run out and vehicles would run out of fuel. Because of the increase of vehicles, the environmental factors and the reduction of gas emissions were also considered. The concepts of sustainability and renewable energy have become closely linked. During the literature review research, some gaps were determined.

First, it is about technology and infrastructure standardization. Electric vehicles are already on our roads, but there are significant differences between jurisdictions. Both vehicles and infrastructural elements still need to be fully standardized.

Second, there needs to be more clarity related to legal issues, uneven regulations, and liabilities regarding autonomous and electric cars. It is challenging to clarify the responsibilities of the central authorities, city and regional administrations, companies that designed, produced, and installed the infrastructure and car manufacturers.

Third, the economic side includes significant investments in infrastructure, car production, logistics, changes in the type of employment and business operations, and personal expenses for purchasing new vehicles.

Fourth, it is necessary to research how customers will accept new technology and regulations, where they see their future benefits, what they consider risks, whether they feel safe, whether their privacy is threatened, and their economic possibilities.

Fifth - planning cities where the most inhabitants will live and creating smart towns where a synergy of mobility and more abundant energy storage will be achieved.

Sixth - predictions of future development and decision-making models based on fuzzy logic should be applied, including all possible scenarios.

Three surveys related to the acceptance of new technology were conducted in 2021 and 2022 - two in Saudi Arabia and one in Australia [9–11]. In addition, a new AV technology adoption framework is presented. In the EV transition we have no ethical and moral issues like with AV transition. Proposed governments regulations and bans on the use of internal combustion engines will accelerate the transition to EV vehicles.

6 Conclusion

The presented study is part of a much broader study based on systematic literature research obtained from the Scopus and WoS databases. The aim of conducting a bibliometric review was to identify the main characteristics of the transition to autonomous and electric vehicles, to follow the evolution of this process and to augment potential trends for future research. This systematic review indicates that more research is needed to identify all challenges of transitioning to autonomous and electric vehicles. In the last five years, advancements in electric over fossil fuel vehicle-share indicated that AVs might use electric powertrain technology as the major one in the future.

Multi-disciplinarity is presented with 54 Scopus research categories. The maturity accomplished by the studies in technical fields, such as engineering, computer science and automation, raised new questions about introducing this technology in the market and the significant impacts and implications of such vehicles on urban transportation.

Results of this study could contribute valuable insights and inputs to the emerging areas of this transition. These areas include public policies and laws, government measures, tax incentives, regulatory aspects, liabilities, cyber-security and hacker attacks, data privacy and security, car safety issues such as the analyses of crashes and accidents, and the probability of accidents with different automation levels applied.

References

1. Elbanhawi, M., Simic, M., Jazar, R.: In the Passenger Seat: Investigating Ride Comfort Measures in Autonomous Cars. IEEE Intell. Transp. Syst. Mag. **7**(3), 4–17 (2015)
2. Taeihagh, A., Lim, H.S.M.: Governing autonomous vehicles: emerging responses for safety, liability, privacy, cybersecurity, and industry risks. Transp. Rev. **39**(1), 103–128 (2019)

3. Elbanhawai, M., Simic, M.: Sampling-Based Robot Motion Planning: A Review. IEEE Access 2014. PP (99)
4. Young, J., Elbanhawi, M., Simic, M.: Developing a Navigation System for Mobile Robots. In: Damiani, E., Howlett, R.J., Jain, L.C., Gallo, L., De Pietro, G. (eds.) Intelligent Interactive Multimedia Systems and Services. SIST, vol. 40, pp. 289–298. Springer, Cham (2015). https://doi.org/10.1007/978-3-319-19830-9_26
5. Elbanhawi, M., Simic, M., Jazar, R.: Receding horizon lateral vehicle control for pure pursuit path tracking. J. Vib. Control 24(3), 619–642 (2018)
6. Elbanhawai, M., Simic, M.: Continuous-Curvature Bounded Trajectory Planning Using Parametric Splines, in Frontiers in Artificial Intelligence and Applications, R.N.-S.e. al., Editor. 2014, IOS Press BV: Chania, Greece
7. Hu, J.Y., et al.: A Decentralized Cluster Formation Containment Framework for Multirobot Systems. IEEE Trans. Rob. 37(6), 1936–1955 (2021)
8. Lim, H.S.M., Taeihagh, A.: Algorithmic Decision-Making in AVs: Understanding Ethical and Technical Concerns for Smart Cities. Sustainability 11(20), 5791 (2019)
9. Todorovic, M., Aldakkhelallah, A.A.A., Simic, M.: The Adoption Appraisal of Autonomous Vehicles. In: International KES Conference on Human Centered Intelligent Systems, KES HCIS 2022, pp. 125–135. Springer, Rhodes, Greece (2022)
10. Abdulaziz Aldakkhelallah, M. Todorovic, and M. Simic, Investigation in Introduction of Autonomous Vehicles. Int. J. Adv. Electron. Comput. Sci. (IJAECS), 8(10), 6 (2021)
11. Abdulaziz Aldakkhelallah, Milan Simic,: Autonomous Vehicles in Intelligent Transportation Systems. In: Alfred Zimmermann, Robert J. Howlett, Lakhmi C. Jain, Rainer Schmidt, (ed.) Human Centred Intelligent Systems: Proceedings of KES-HCIS 2021 Conference, pp. 185–198. Springer Singapore, Singapore (2021). https://doi.org/10.1007/978-981-16-3264-8_18
12. Page, M.J., et al.: The PRISMA 2020 statement: an updated guideline for reporting systematic reviews. Syst. Rev. 10(1) (2021)
13. Sovacool, B.K., Axsen, J.: Functional, symbolic, and societal frames for automobility: Implications for sustainability transitions. Transp. Res. Part a-Policy Pract. 118, 730–746 (2018)
14. Wu, J.W., Liao, H., Wang, J.W.: Analysis of consumer attitudes towards autonomous, connected, and electric vehicles: a survey in China. Res. Transp. Econ. 80, 100828 (2020)
15. Muller, J.M., Comparing technology acceptance for autonomous vehicles, battery electric vehicles, and car sharing-A Study across Europe, China, and North America. Sustainability 11(16), 4333 (2019)
16. Weiss, J., Hledik, R., Lueken, R., Lee, T., Gorman, W.: The electrification accelerator: Understanding the implications of autonomous vehicles for electric utilities. Electr. J. 30(10), 50–57 (2017)
17. Dlugosch, O., Brandt, T., Neumann, D.: Combining analytics and simulation methods to assess the impact of shared, autonomous electric vehicles on sustainable urban mobility. Inform. Manage. 59(5), 103285 (2022)
18. Hou, F.X., et al.: Comprehensive analysis method of determining global long-term GHG mitigation potential of passenger battery electric vehicles. J. Cleaner Prod. 289, 125137 (2021)
19. Todorovic, M., Simic, M., Kumar, A.: Managing transition to electrical and autonomous vehicles. Knowl.-Based Intell. Inform. Eng. Syst. 112, 2335–2344 (2017)
20. Mazur, C., et al.: Comparing the effects of vehicle automation, policymaking and changed user preferences on the uptake of electric cars and emissions from transport. Sustainability 10(3) 676 (2018)
21. Prado, A.J., Michalek, M.M., Cheein, F.A.: Machine-learning based approaches for self-tuning trajectory tracking controllers under terrain changes in repetitive tasks. Eng. Appl. Artif. Intell. 67, 63–80 (2018)

22. Hajebrahimi, A., Kamwa, I., Huneault, M.: A novel approach for plug-in electric vehicle planning and electricity load management in presence of a clean disruptive technology. Energy **158**, 975–985 (2018)
23. Todorovic, M., Simic, M.: Managing transition to autonomous vehicles using bayesian fuzzy logic. Smart Innov., Syst. Technol. **145**, 409–421 (2019)

An Investigation in Autonomous Vehicles Acceptance

Abdulaziz Ayedh A. Aldakkhelallah$^{(\boxtimes)}$ ⓘ, Milan Todorovicⓘ, and Milan Simicⓘ

RMIT University, Melbourne, VIC 3000, Australia
Abdulaziz.adk@gmail.com

Abstract. Future road transport will significantly be impacted by the arrival of autonomous vehicles. Self-driving cars provide a chance to cut hazardous emissions and promote a more sustainable future, through more efficient and faster traffic solutions. Apart from those benefits, there are concerns such as safety, security, and the cost of the transport system transition. There are also questions about the costs and models of car ownership, which are possible to introduce thanks to the automation. Like with any new technology customer and public acceptance are key factors. By performing longitudinal study, thorough surveys, in different communities and over a long time period, we intended to investigate possible adoption and integration of autonomous vehicles into the communities. In order to get better insight into the degree of readiness for this new technology, among both the general public and enterprises, the results of the survey are analysed and presented in this document. The results of our investigation will be helpful in determining if and when autonomous vehicles are likely to become a mainstream form of transport in the future and what role they will play in building, sustainable transport solutions, smart cities and communities. This is just one of the future several observations on the same topic of AV introduction, that will be lasting for few years till the technology is finally introduced.

Keywords: Survey · Autonomous Vehicles · Transitions · Smart Cities · Technology Acceptance

1 Introduction

Two major technology changes in automotive industry are occurring now: electrification and automation. The vision of the car of the future is an electrical autonomous vehicle. We will have wide range of hybrid solutions and we already have many of those now. For a long time, we are conducting research in booth new electrical vehicles (EV) solutions [1–3] and autonomous vehicles (AV) [4–8]. While engineering is one side of the story public acceptance of any new technology is also crucial. Following that, we also conduct longitudinal research in the management of those two transitions [9–14]. During 2021 and 2022 research team from RMIT University, School of Engineering, has conducted investigation in the introduction of the new AV technologies. In reference to AV introduction, we have conducted public surveys in Saudi Arabia and Australia. We

© The Author(s), under exclusive license to Springer Nature Singapore Pte Ltd. 2023
A. Zimmermann et al. (Eds.): KES-HCIS 2023, SIST 359, pp. 78–87, 2023.
https://doi.org/10.1007/978-981-99-3424-9_8

have already published our findings from KSA [15–17] and now we present analysis of the currently available data from Australia 2023 survey. Statistically, a minimum sample size to get meaningful results is 100. Since we are doing longitudinal study, we will have more reliable, i.e., more statistically meaningful data along the timeline, till the full introduction of autonomous vehicles worldwide.

1.1 Levels of Autonomy in Driving

The degree to which a machine or system can function autonomously without human intervention is referred to as its level of autonomy. According to the Society of Automotive Engineers' (SAE) definition of autonomy levels, given in the SAE J3016 standard, there are currently six categories. They go from complete autonomy (level 5) to no autonomy (level 0), with varying degrees of decision-making and control given to the machine. A dynamic driving task (DDT) procedure that incorporates operational, tactical, and strategic duties identifies each level. Levels have different DDT fallback mechanisms, and for levels 1 through 3, the driver is in charge of carrying out the fallback procedure. The system is responsible for the operation of the DDT fallback mechanism at levels 4 and 5. The automatic driving system (ADS) in level 5 is in control of DDT performance and, if necessary, performs the DDT fallback procedure. All levels of automation will coexist for a considerable amount of time as the integration of automation in the automobile sector is a gradual process. Driver control of the cars is necessary for the first three levels of autonomy, which are now in widespread use. A good review of the research literature in autonomous technology is given here [18].

2 Technology Acceptance Framework

When a new technology is made available to consumers, the life cycle of that technology begins. Figure 1 shows technology acceptance model for autonomous vehicles, i.e., how are AVs going to be introduced. The shift to AVs must ensure traffic safety, data security and no ethical, moral, or legal uncertainties. Based on data collecting and reaction algorithms, vehicles with lower degrees of autonomy (below levels 4 and 5) already have practical and affordable driver assistance solutions and active safety, like lane change warning, blind spot warning, and parking assistance [19].

   The introduction of AVs, from the technical point, which is referred to as the engineering state in Fig. 1, is viewed as technologically dependent on the development of vision, safety, path planning, localisation and mapping. It is also including Inter-Vehicle Communication (IVC) systems and vehicular ad hoc network (VANET) [20]. The other three states, on the other hand, indicate legal, the moral and ethical issues that the introduction of AVs also brings up in. These enquiries are a result of the coexistence of AVs and non-AVs as well as the accountability for AV-related incidents. Manufacturers, authorities, and users must come to agreements and final solutions. Different communities, throughout the world, may have different views on how crucial road circumstances should be handled, resulting in various software modules that are specialised to particular regions, or nations. This is comparable to the distinction between countries with left and right-hand driving. There is currently no acknowledged moral or ethical code

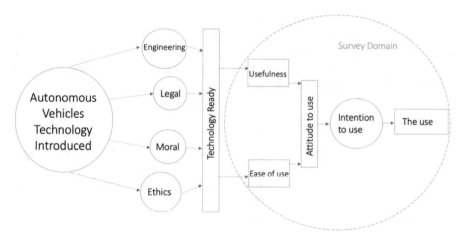

Fig. 1. AV technology acceptance model and our survey domain in green

that applies to everyone in the globe. A large survey of 1533 persons was conducted in U.S., the U.K., and Australia [21]. In Fig. 1, the fourth base state stands in for legal issues, which differ both internationally and inside big nations, where each state has its own laws. Vehicle and roadside data, including videography and personal data, must be protected, and handled morally, ethically, and lawfully. Countries worldwide establish guidelines for secure and transparent data processing. Each nation is required to create regulations based on its national laws.

Customers will determine if they wish to employ a technology when it is ready for use, as seen in the right-hand side of Fig. 1. The outcomes demonstrated that the advantages of the technology and its simplicity of use were key factors in its acceptance, while the price is slowing that process down. The demand and likelihood of adoption of a technology rise with its utility, dependability, safety, and simplicity. Most of engineering questions about the new AV technology currently have a satisfactory solutions, but the legal, moral, and ethical issues are still unsolved and have variable degrees of resolution.

3   Survey Data Analysis

For our survey we have used Qualtrics software. It is a web-based application for developing and analysing surveys. The Australian AV survey was approved by the RMIT University Ethics Committee (#23507) in 2021. All mandatory instructions and introduction were provided to participants at the beginning. Currently, in this survey, 117 volunteers of all ages, educational levels, driving preferences, and key stakeholder positions participated in the study. We had 108 usable responses. No personal data was gathered or kept. Participants could exit at any moment and did not have to answer all questions. Following that, we have useful number of responses for particular questions in the range of 105 to 108. Survey software is giving the count and the percentage for all answers taking that particular number for each question answered separately.

In order to learn more about their attitudes towards AVs and acceptance of new technology we have conducted survey in the domain as shown in Fig. 1. The purpose of the poll was to explore public interest in using AVs and to better understand consumer perceptions. Participants included both regular drivers and stakeholders, and they represented all ages, driving styles, educational levels, and occupations. Although personal vehicles are used extensively in Australia, the introduction of public transport and car-sharing, particularly with autonomous vehicles, is becoming a trend as it is internationally, as well.

First of all, we have asked questions that helped us to establish demographics picture of participants, as shown in the Table 1: gender, the age, education level, current mode of transport used, stakeholders' groups and management role in the organisation. Then we enquired about familiarity with new technology and seen car ownership in the future. We also asked about seen benefits and concerns, or risks in relation to new technology. There are questions about pollution reduction and efficiency and sustainability in the transport. Following that they have given us their vision, i.e., opinion when the transition will happen. There are 34 questions in total.

We are investigating the connection between participant attributes and technological acceptability. As an illustration, we divided the participants' ages into the following ranges: 18–25, 26–35, 36–55, and over 55. The general hypothesis is that people who are young, between 26 to 35, are more open to new technology. The price might also be a concern to all generations as we can see later from other graphs (Fig. 2).

Every group has a different perspective on emerging technologies. Figure 3 shows correlation between two variables: expected benefits of the transition and the educational level. There is no linear coloration between educational level and seen benefits. This means that our surveyed population has enough knowledge and interest in new technology regardless of being with high school of postgraduate degree. They are all educated and ready for the new technology, up to the different degrees. In conclusion, these findings show that a sizeable portion of participants think the switch to new technology would be advantageous (Fig. 4).

The management levels define the ranking of people who work in an organisation.

At the level 2 management role are usually supervisors in the big corporation. They choose the importance of comfort on the future autonomous vehicles. This may mean that they are hardworking while supervising teams and need some comfort while driving home. High level managers are less concerned about the comfort of drive-less cars as they often have human drivers. This is just one of the hypotheses based on our data. No management role, which is usually played by the ordinary employee in the company, do not care about the comfort on future transport. This is another hypothesis. All of the hypotheses will be further explored with new surveys, i.e., out longitudinal approach that will include large number of participants, over time and conducted nationally.

How stakeholders see the transition and the extent of their support for the change is another critical viewpoint. Table 2 gives an example of this. Representatives of the business community, government agencies, traffic authorities, and the general public are important stakeholder groups. Industry and the general public act as consumers while important parties accountable for the transformation include the government and traffic authority.

Government participants made up to 20% of the sample, while industrial participants made up 36.4%. Participants from traffic authorities made up the majority, 100%, while

Table 1. Demographics of the survey participants.

Demographic	n	%
Gender		
Male	85	80
Female	20	20
Age group		
18–25	11	10
26–35	51	48
36–50	28	26
50<	17	16
Stakeholder group		
Industry	22	20
Government	15	14
Traffic Authority	1	1
Just Public	70	65
Educational Level		
Primary school	0	0
High school	6	6
Diploma	10	9
Graduate degree	25	23
Postgraduate degree	67	62
Management Role		
Level 5	13	12
Level 4	23	21
Level 3	22	21
Level 2	10	9
Level 1	6	6
No Management	33	31
Transport mode		
Car owner	82	76
Public transport	26	42

members of the general public made up 11.4%. We do not have enough data for traffic authorities group. The findings indicated that participants from the industry sector were less concerned about the adoption of driverless vehicles in than were traffic officials.

Due to the high cost of manufacturing, research and development, new technologies are frequently introduced at a high cost. In its life cycle, this new technology goes through several stages, including Introduction, Growth, Maturity, and Decline. Figure 5 depicts the findings, which reveal that 34% of the participants were concerned about the high cost of autonomous vehicles relative to the advantages it offers. This has indicated how everybody think about the importance of the price. This is one of the explanations

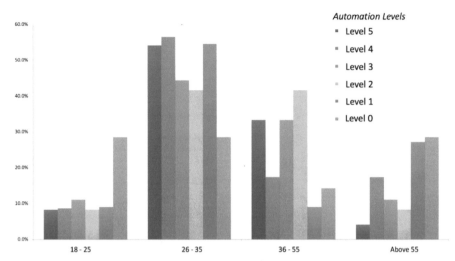

Fig. 2. Ready to buy status as function of the age of the participants. Age group of 26–34 are ready to invest to technology, giving the priority to level 4.

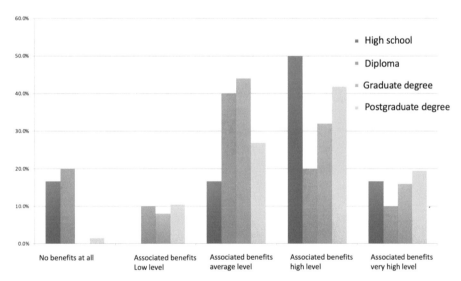

Fig. 3. Benefits from the transition as seen by the participants with different educational level.

for the introduction of new ownership models for autonomous vehicles, such as shared ownership.

Based on Fig. 6, transition will happen between 2030, and 2040, say 2035 as average of two highest counts. As we have conducted 2 surveys in Saudi Arabia, we have found that in KSA expectation is 2030.

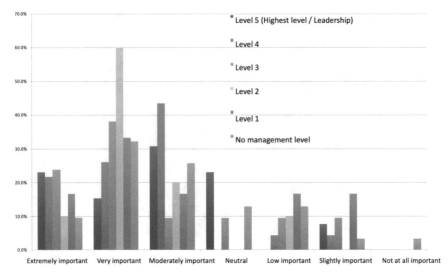

Fig. 4. Correlation between management level and the comfort associated with autonomy level. Level 2 management are more exciting to have comfort autonomous vehicles.

Table 2. The correlation between stakeholders' groups and the seen benefits.

Q12: 12) T...vehicles?	Q5: 5) Which of the key stakeholders' groups do you belong to?			
	a) Industry	b) Government	c) Traffic authoriti...	d) Just public
a) No benefits at all	4.5%	6.7%	0.0%	2.9%
b) Low level	13.6%	0.0%	0.0%	10.0%
c) Average level	13.6%	40.0%	0.0%	35.7%
d) High level	31.8%	33.3%	100.0%	40.0%
e) Very High level	36.4%	20.0%	0.0%	11.4%
Total	100.0%	100.0%	100.0%	100.0%

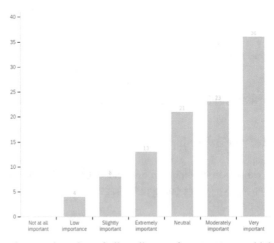

Fig. 5. Answers to the question: Out of all attributes of autonomous vehicles how important is the price?

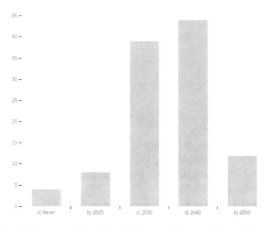

Fig. 6. Expected timeline of the transition by the participants.

4   Conclusion

A new technology introduction framework flowchart shows how autonomous vehicles go from their introduction phase to their growth phase. Although there are global and local legal, moral, and ethical issues that need to be resolved, communities are preparing to use this new technology. We wanted to know if the general public and other important stakeholders are ready for this transformation. Modern automotive engineering, information, and communication technologies, VANET, the Internet of Things, cloud computing, and artificial intelligence are all available and used in the automotive technology. However, there are still a number of issues, from the soft systems approach in management, that need to be resolved. Governments and traffic authorities must create regulations, communities must agree on moral and ethical issues, and software developers must be tasked with incorporating these conclusions into the AI in onboard computers. Those are all research directions and actions for all stakeholders. Every community, countries and their governments and authorities must follow those directions and all of that is happening according to the given technology acceptance framework, presented in the paper.

According to our findings majority of respondents had previous knowledge of autonomous vehicles, have positive attitudes to new technology, and high expectations about the benefits. At the same time, they are aware of safety and security risks and of course the high price at this stage of development.

According to our surveys, conducted in two countries and longitudinally, we conclude that the public believes that smart cities and autonomous vehicles would be fully developed in 7 to 15 years, which fits nicely with the project's suggested timeframe. Although autonomous vehicles are only a small part of the overall Intelligent Transport Systems and smart cities, it is reassuring that the projected project timeframe and the anticipated transition have a fair timeline association.

Acknowledgment. Author Abdulaziz Aldakkhelallah acknowledges the financial support of the Saudi Arabia government under the Custodian of the Two Holy Mosques Scholarship Program - King Salman Scholarship Program.

References

1. Yulianto, A., et al.: Modelling of full electric and hybrid electric fuel cells buses. Procedia Comput. Sci. **112**, 1916–1925 (2017)
2. Royale, A., Simic, M., Lappas, P.: Engine exhaust manifold with thermoelectric generator unit. Int. J. Engine Res., 1468087420932779 (2020)
3. Royale, A., Simic, M.: Research in vehicles with thermal energy recovery systems. Procedia Comput. Sci. **60**, 1443–1452 (2015)
4. Elbanhawi, M., Simic, M., Jazar, R.: Receding horizon lateral vehicle control for pure pursuit path tracking. J. Vib. Control **24**(3), 619–642 (2018)
5. Elbanhawi, M., Simic, M., Jazar, R.: Solutions for path planning using spline parameterization. In: Jazar, R.N., Dai, L. (eds.) Nonlinear Approaches in Engineering Applications, pp. 369–399. Springer, Cham (2016). https://doi.org/10.1007/978-3-319-27055-5_13
6. Young, J., Elbanhawi, M., Simic, M.: Developing a navigation system for mobile robots. In: Damiani, E., Howlett, R.J., Jain, L.C., Gallo, L., De Pietro, G. (eds.) Intelligent Interactive Multimedia Systems and Services. SIST, vol. 40, pp. 289–298. Springer, Cham (2015). https://doi.org/10.1007/978-3-319-19830-9_26
7. Elbanhawi, M., Simic, M., Jazar, R.: In the passenger seat: investigating ride comfort measures in autonomous cars. IEEE Intell. Transp. Syst. Mag. **7**(3), 4–17 (2015)
8. Elbanhawi, M., Simic, M.: Sampling-based robot motion planning: a review. IEEE Access **2**, 56–77 (2014)
9. Aldakkhelallah, A., Todorovic, M., Simic, M.: Investigation in Introduction of Autonomous Vehicles. International Journal of Advances in Electronics and Computer Science (IJAECS) **8**(10), 6 (2021)
10. Todorovic, M. and M. Simic. *Transition to Electrical Vehicles Based on Multi-Attribute Decision Making.* in *2019 IEEE International Conference on Industrial Technology (ICIT).* 2019
11. Todorovic, M., Simic, M.: Managing transition to autonomous vehicles using bayesian fuzzy logic. In: Chen, Y.-W., Zimmermann, A., Howlett, R.J., Jain, L.C. (eds.) Innovation in Medicine and Healthcare Systems, and Multimedia: Proceedings of KES-InMed-19 and KES-IIMSS-19 Conferences, pp. 409–421. Springer Singapore, Singapore (2019). https://doi.org/10.1007/978-981-13-8566-7_38
12. Todorovic, M., Simic, M.: Feasibility study on green transportation. Energy Procedia **160**, 534–541 (2019)
13. Todorovic, M., Simic, M.: Current state of the transition to electrical vehicles. In: De Pietro, G., Gallo, L., Howlett, R.J., Jain, L.C., Vlacic, L. (eds.) KES-IIMSS-18 2018. SIST, vol. 98, pp. 130–139. Springer, Cham (2019). https://doi.org/10.1007/978-3-319-92231-7_14
14. Todorovic, M., Simic, M., Kumar, A.: Managing transition to electrical and autonomous vehicles. Procedia Comput. Sci. **112**, 2335–2344 (2017)
15. Todorovic, M., Abdulaziz, A.A., Aldakkhelallah, M.S.: The Adoption Appraisal of Autonomous Vehicles. In: Zimmermann, A., Howlett, R.J., Jain, L.C. (eds.) Human Centred Intelligent Systems: Proceedings of KES-HCIS 2022 Conference, pp. 125–135. Springer Nature Singapore, Singapore (2022). https://doi.org/10.1007/978-981-19-3455-1_9
16. Aldakkhelallah, A.A.A., et al.: Public opinion survey on the development of an intelligent transport system: a case study in Saudi Arabia. AIP Conf. Proc. **2681**(1), 020089 (2022)

17. Aldakkhelallah, A., Todorovic, M., Simic, M.: Investigation on the acceptance of autonomous vehicles. In: Autonomous Vehicle Technology Conference - APAC21, 21st Asia Pacific Automotive Engineering Conference. SAE Australia, Melbourne (2022)
18. Khan, M.A., et al.: Level-5 autonomous driving—are we there yet? a review of research literature. ACM Comput. Surv. **55**(2), 27 (2022)
19. Simic, M.: Vehicle and public safety through driver assistance applications. In: Wellnitz, J., Subic, A., Leary, M. (eds.) Sustainable Automotive Technologies 2010, pp. 281–287. Springer, Heidelberg (2009). https://doi.org/10.1007/978-3-642-10798-6_34
20. Simic, M.: Vehicular Ad hoc networks in telecommunication in modern satellite, cable and broadcasting services (TELSIKS). In: 2013 11th International Conference on 2013. IEEE Conference Publications, Nis (2013)
21. Schoettle, B., Sivak, M.: A survey of public opinion about autonomous and self-driving vehicles in the U.S., the U.K., and Australia. In: The University of Michigan Transportation Research Institute Michigan, p. 42 (2014)

Edge Computing Technologies for Mobile Computing and Internet of Things (3rd Edition)

Can Business Be Sustainable: A Case Study of the Information Technology Sector

Soukaina El Maachi[1]([✉]), Rachid Saadane[1], and Abdellah Chehri[2]

[1] SIRC/LAGES-EHTP, Hassania School of Public Works, Casablanca, Morocco
{soukaina.elmaachi.cedoc,saadane}@ehtp.ac.ma
[2] Department of Mathematics and Computer Science, Royal Military College, Kingston, Canada
chehri@rmc.ca

Abstract. There is a worldwide climate emergency. As the world changes, new technologies have the potential to help us get closer to a more sustainable future. But what should business executives know about these technologies? Driven rather by short-term gain, than long-term uncertainty, leaders must be aware of the options available and how they might be incorporated into their operations as sustainable technology becomes more prominent. To incorporate sustainability into their operations, businesses must go beyond volunteer social and environmental activities. This paper explores how companies can pursue economic growth while meeting long-term sustainability goals such as carbon neutrality. We discuss some conceptual frameworks proposed by global initiatives to evaluate the potential sustainability of business models. Furthermore, we analyse the information technology sector and how businesses operating in or closely associated with it can reduce their carbon footprint and mitigate the effects of the business, in general, and technology, in particular, on the climate. Finally, we discuss how companies could use AI to make their businesses more sustainable.

Keywords: Sustainability · Internet of Things · Industry 4.0 · Information Technology · Climate Change

1 Introduction

There is a climate emergency on a worldwide scale, atmospheric pollution levels are at an all-time high, and the temperature over the previous few years has been the highest it has ever been recorded. As time goes on and the world changes, emerging technologies may be able to help us get one step closer to an environmentally friendly future. Yet, what exactly is it that top executives in businesses need to know about these technologies? In order to successfully implement them into their corporate operations and goals, companies will need to make informed judgments [1].

Since the Meadows report of 1972, the Charney report on the climate of 1979, the countless summits on the environment, and the recent reports of the intergovernmental panel on climate change, the discourse that maintains the belief that technological advancements will save us from the effects of global warming remain widespread, if not

© The Author(s), under exclusive license to Springer Nature Singapore Pte Ltd. 2023
A. Zimmermann et al. (Eds.): KES-HCIS 2023, SIST 359, pp. 91–100, 2023.
https://doi.org/10.1007/978-981-99-3424-9_9

predominate. In other words, there would be no need for us to change our behaviors that require a significant amount of energy in terms of our carbon footprint or the extraction of resources that are necessary to manufacture our technologies.

As sustainable technology becomes more prevalent, leaders need to be aware of the available options and how they might be incorporated into their operations in order to maximize short-term gain while minimizing long-term uncertainty. This awareness should be driven more by the potential for short-term growth. If a company is serious about incorporating sustainability into its business, it must go above and beyond merely participating in volunteer social and environmental activities.

This article provides some thoughts regarding how businesses might pursue economic growth while keeping up with their long-term sustainable aims to reach carbon neutrality. We will analyze some conceptual frameworks that have been put forward from worldwide initiatives to assess the potential sustainability of business models. These frameworks will be discussed concerning the possible sustainability of business models.

We discussed the information technology industry and how companies that are either directly involved in this industry or are closely connected to it can lessen their impact on the environment by lowering their carbon footprint and mitigating the negative effects that both businesses, in general, and technology, in particular, has on the climate. In the previous and final segment, we covered how organizations can implement AI to make their operations more environmentally friendly.

2 Sustainable Development

There are currently no widely used indicator sets related to sustainable development that is supported by a strong theory, subject to meticulous data collection and analysis, and have significant policy impact. The authors of [4] provide three main justifications:

1. Confusing terminology, data, and measurement methods.
2. The ambiguity of long-term development.
3. The variety of goals in defining and measuring sustainable development [8].

While there is still some ambiguity surrounding the definition of sustainable development, it is gradually being resolved. Global and local consensus is increasingly adopting goals and targets for sustainable development. The Bruntland Commission presented its report "Our Common Future" in 1987 to connect economic growth and environmental stability concerns. This report defined sustainable development as "development that meets the needs of the present without compromising the ability of future generations to meet their own needs" [5].

As a result of increasing advancements in environmental monitoring and data reporting, the Environmental Performance Index for 2022 will be able to include various cutting-edge indicators. To complement the ongoing conversations about climate policy, this index offers a new indicator that projects nations' progress toward reaching net-zero emissions in the year 2050. The net-zero in 2050 metric can be used by policymakers, the media, business executives, non-governmental organizations, and the general public to assess the effectiveness of national policies, identify the major contributors to climate change, and mobilize support to correct off-track emitters' emissions trajectories [6].

In 1995, the Tellus Institute and the Stockholm Environment Institute collaborated to found the Global Scenario Group to assemble an influential and multiethnic group of individuals from all over the world to investigate the potential outcomes of global development in the twenty-first century. In the years that followed, the scenario framework and quantitative analysis presented by the GSG were utilized in a significant number of studies on a global, regional, and national scale [7].

Table 1. Environmental Indicators and Targets

Region	Indicator	1995	2025	2050
Climate				
World	CO_2 concentration	360 ppmv	Stabilize at <450 ppmv by 2100	
OECD	CO_2 emissions rate	Various and rising	Increases slowing, energy efficiency rising	<35% of 1990
Non-OECD	CO_2 emissions rate	Various and rising	Reach OECD per capita rate by 2075	
Resource Use				
OECD	Eco-efficiency	$100 GDP/300 kg	$100 GDP/75 kg	$100 GDP/30 kg
non-OECD	Eco-efficiency	Various but low Various but low	Converges towards OECD practices	
Toxics				
OECD	Releases of persistent organic pollutants	Various but high	<50% of 1995	<10% of 1995
non-OECD	Releases of persistent organic pollutants	Various and rising	Increases slowing	Converges to OECD per capita values
Frechwater				
World	Use-to-Resource ratio	Various and rising	Reaches peak values	0.2–0.4 maximum
	Population in water stress	1.9 billion	Less than 3 billion	Less than 3.5 billion
Ecosystem Pressure				
World	Deforestation Land degradation Marine over-fishing	Varius but high Varius but high Fishstock declining	No further deforestation	Net reforestation Net restoration Healthy fish stoc

The Global Scenario Group uses 65 indicators to define global equality, national equity, hunger, energy consumption, water use, deforestation, carbon emissions, sulfur emissions, and toxic waste. Some of these issues include global equity and national equity [8]. Table 1 presents these indicators for a total of five different environmental concerns.

2.1 Climate

Long-term goals for the climate are outlined in the Framework Convention on Climate Change, and one of those goals is to maintain stable amounts of greenhouse gases in the atmosphere. In this case, the criterion that has been decided upon is that between 1990 and 2100, there should be average warming at a pace of no more than 0.1 °C per decade; this will provide many ecosystems the opportunity to readjust [10].

2.2 Resource Use

The extraction, refining, manufacture, transportation, and ultimate disposal of resources substantially contribute to environmental degradation and waste in modern economies. These activities also contribute to the creation of new resources. These materials include, for example, the metals and chemicals that are utilized in the automotive industry, the chemicals that are utilized in the production of paints, pesticides, and an infinite number of other items, as well as minerals and fibers [9].

2.3 Toxic Substances

The widespread industrial release of dangerous substances, such as heavy metals and persistent organic pollutants, is a significant issue. The objective outlined below presupposes a reduction of 50 percent of all emissions, discharges, and losses in OECD (Organization for Economic Cooperation and Development) countries by the year 2025 and a reduction of 90 percent by the year 2050 [9].

2.4 Freshwater

The maintenance of economic activity, the fulfillment of human needs, and the protection of ecosystems all require the presence of freshwater. In principle, one reasonable goal for achieving sustainability would be to lower water pressure over the following decades in all locations where there is a risk that a lack of water could impede growth or cause damage to ecosystems.

The use-to-resource ratio values are aimed to be maintained at a level between 0.2 and 0.4 by the year 2050. Freshwater withdrawals in 2050 should be lower than those in 1995 for countries whose use-to-resource ratio exceeds 0.4. These countries are primarily located in the Middle East and North Africa [3].

2.5 Ecosystem Pressure

Increasing demands are putting a strain on natural ecosystems to provide sufficient food and other resources. The problem is made even more complex by the expansion of constructed environments created to fulfill the needs of growing populations and economies in terms of housing, business, and transportation. It is imperative that fragile ecosystems be protected, that destructive logging practices be put an end to, that arid lands be carefully managed to prevent desertification, that tilled lands be carefully managed to avoid erosion or other forms of degradation, and that the expansion of built-up areas be moderated in order to avoid severe damage [9].

3   Business Sustainability

The authors of [14] conducted research into the relationship between small enterprises and environmentally friendly development. The result has demonstrated that: small business managers are concerned about sustainable development; nonetheless, financial concerns are the primary drivers of their activity. Small businesses are willing to participate in activities that contribute to sustainable development, yet, they seek instant gratification and anticipate financial rewards.

3.1   New Vision of Business Drivers

Sustainable use of natural resources cannot be ensured by how society's players are currently arranged or interacting. The signs of this serious scenario have been around for a while, and there needs to be more effort or progress toward global sustainable development.

Corporations, who may rely on technological and financial ability while carrying an institutional function to contribute to global sustainable development, are one of the actors accountable for this predicament. Therefore, despite disagreements over the definitions of sustainability and sustainable development, "corporate sustainability" describes a firm's ability to contribute to global sustainable development and all the issues related to how the economy, society, and environment are interconnected.

Promoting shared values offers businesses the chance to succeed while directly contributing to society's advancement. Innovation is essential to help companies to transition from conventional business models to sustainable ones [11].

3.2   Conceptual Frameworks

Many initiatives have been launched to assess the impact of corporate sustainability goals on their bottom lines. In [2], a comprehensive and integrative performance assessment framework was used to promote the identification of sustainability innovations. The research looked at the framework's benefits and limitations.

The business models of four sustainability leaders were examined using the proposed methodology: (1) Unilever, (2) Kao Group, (3) Woolworths, and (4) Lotte Shopping.

The study shows that when attempting to promote long-term value from a business, the performance dimensions of stakeholder satisfaction, strategic drivers, business processes, capabilities, and stakeholder contributions all contribute significantly to a better understanding of the firms' business models.

According to Fig. 1, the sustainability of a company's value proposition is more closely related to how well it performs in terms of sustainability when it comes to the satisfaction of its stakeholders and corporate strategic drivers. Both elements clearly state who and how the company wishes to advance its value.

According to [12], modern organizations are increasingly adopting ethical behaviors as they pursue profitable endeavors. A thorough literature analysis indicates a connection between corporate social performance (CSP), often known as CSR, and financial performance.

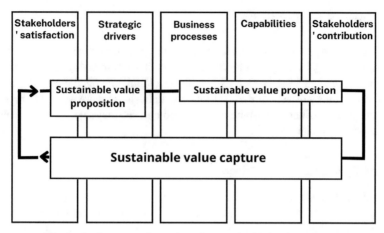

Fig. 1. Performance dimensions for sustainable business model.

Corporate social responsibility (CSR) has consistently been contested by those who want businesses to go beyond stakeholder involvement, ethics, and transparency. New sustainable business models created to address environmental, socioeconomic, and governance inadequacies are increasing, including ethical behavior.

3.3 Benefits of Incorporating Sustainability in Businesses

- Serious climate action could boost the U.S. GDP by $3 trillion by 2070. Almost 80% of businesses now report on sustainability, up from 12% in 1993. Assets managed by ESG funds in the financial markets increased to $330 billion in 2021.
- Successful businesses approach sustainability from the perspective of how they can use it to improve their capital returns operationally and economically. When businesses utilize the appropriate approaches to achieve their sustainability goals, they uncover cost and supply chain savings, can frequently lower packaging or service prices, and, in far too many cases, find creative ways to boost profits on their green goods [12].
- These kinds of activities can benefit a brand's reputation in the marketplace, the growth of its customer base, and customer loyalty [12].
- A company's perspective on sustainability should be categorized as a strength and, ideally, a differentiator among its competitors when performing a standard SWOT analysis. When consumers weigh their options, it can even be the determining factor in their decision.

4 Technological Business Sustainability

According to [15], the information technology revolution is expected to bring about economic changes that are almost as significant as the industrial revolution itself. The term "carbon footprint" has long been used to refer to a product's or service's life-cycle carbon equivalent emissions and consequences [15].

Although the modification, storage, and exchange of digital data may appear to have no environmental impact, several cases appear to show otherwise. High-speed, high-bandwidth connectivity between our homes and offices may allow us to work remotely. Still, it may also exacerbate urban expansion if people begin to live further away from their places of employment. E-commerce product packaging for shipping may require more resources and energy than product packaging for in-store purchases.

4.1 How Technology Contributes to Unsustainable Development

Technology production and use can pollute the air, water, heat, and noise:

Nonrenewable resources, such as gold and other precious metals, are used to manufacture technology. Many others, including coal, are used to generate the electricity required by the technology. Even renewable resources, such as trees and water, are being poisoned or depleted faster than they can be replenished [13].

– Manufacturing technology generates a large amount of waste, and broken or outdated devices and computers are discarded when they are no longer functional. These electronics, also known as "techno trash," contain hazardous substances harmful to the environment. They necessitate specific disposal methods [21].
– There are approximately 22 billion Internet-connected devices on the planet, many of which are frequently updated. According to the United Nations Environment Program, this generates around 50 million tons of technological waste annually [17].
– Currently, the Information and Communications Technology (ICT) sector accounts for more than 2% of global emissions. Nonetheless, if current trends continue, it will account for 15% of emissions by 2040, or half of the global transportation emissions. The rapid development of the Internet of Things (IoT) is one factor in this (IoT) [16, 22].
– Juniper Research predicts that by 2024, there will be 83 billion IoT connections, up from 35 billion in 2020—a 130% increase in just four years. Furthermore, by 2023, 66% of the world's population will have access to the internet, up from 51% in 2018 [16]. Is there going to be enough electricity to power all of these devices as more people connect? [23]
– The increase in the number of devices and cloud-based services is driving the growth of data centers, which consume 2% of the world's electricity. By 2030, that statistic could be as high as 8%.
– The use of potentially carcinogenic substances, which can cause cancer, as well as technology addiction, can lead to additional health problems such as obesity and carpal tunnel syndrome.

4.2 Solutions to Alleviate the Impacts of Technology on the Climate

Understanding how the ICT sector affects the environment may be difficult because it is intangible. The information and communications technology industry is rapidly expanding, accounting for more than 3% of global emissions, or roughly the same as the aviation sector's fuel-related carbon footprint. The environmental impact has become too great to ignore, with factors such as the proliferation of devices, the rise of data centers, and the ambiguity of carbon offsets all contributing to the problem. The good news is that with careful investigation and planning, we can work towards solutions to begin a shift.

- Carbon offset purchases by businesses are becoming increasingly popular to mitigate the negative consequences of their carbon emissions.

- Some data centres (such as AWS) will have established five carbon-neutral zones by 2020, for which they will purchase carbon offsets to offset the emissions generated in those areas.

- A carbon offset is a reduction in greenhouse gas emissions, such as carbon dioxide emissions, that is used to compensate for emissions generated elsewhere. When someone buys a carbon offset, their money is used to fund a project that reduces or prevents new greenhouse gas emissions. The most common carbon offset programs are initiatives to manage forests or generate electricity from renewable sources.

- System optimization (for example, through Artificial Intelligence).
- Locating data centers in colder climates to reduce the cost of cooling [17].
- Using renewable energy - Increasing the average lifespan of gadgets and recycling those that are discarded, which will be exacerbated by the impending introduction of 5G and its increased capacity to connect things to the Internet.
- The European Parliament adopted the "Longer Product Lifetime: Benefits for Consumers and Businesses" decision in 2017. This regulation aims to reduce the number of scrap parts in technology by making more tools available for consumers to repair their equipment and providing tax breaks to businesses that make their products more durable.
- Businesses must encourage recycling: According to the United Nations (UN), only 20% of garbage is recycled. We must "urgently shrink our ecological footprint by changing how we create and consume goods and resources," according to Sustainable Development Goal of the UN [17].

5   How Artificial Intelligence Can Help Businesses Achieve Sustainability

5.1   Using AI to Reduce Waste

Leading Italian multi-utility Hera SpA is investigating how artificial intelligence (AI) can reduce landfill waste by directing more recyclable material towards environmentally preferable outcomes. As Italy's largest waste management and recycling company and a provider of electricity, water cycle management, and heating services, Hera is at the forefront of the ongoing crisis to reduce waste and environmental harm. The company is known for being innovative, combining ecological responsibility with a forward-thinking attitude [18].

Hera's staff performs manual trash analysis. Spotters look for recoverable materials such as plastics, glass, metal, and organic material as trash is pushed towards conveyors when trucks unload at the plants' entrance and help guide downstream sorting. Consider how difficult the job becomes when there are 1,400 spotters spread across 89 factories. 6.3 million tons of trash are handled each year. In other words, there is significant room for improvement. The goal is to capture footage of incoming trash and train AI to recognize features of objects and materials that allow them to be recovered and reused.

Hera and the IBM Garage team collaborated to create and publish a minimum viable product (MVP) that employs IBM Watson® Studio and IBM Watson Machine Learning technologies to create a one-of-a-kind solution for the use case and includes a machine learning model to identify the primary waste patterns [18].

5.2   Using AI to Improve Quality Control

One way AI alters manufacturing procedures is through highly precise quality control. Over time, the AI learns about the manufacturing process by observing how each piece of machinery works. It can quickly detect flaws, other problems in computers, and all irregularities.

5.3   Using Artificial Intelligence to Improve Digital Twin Technology

A digital twin is a precise virtual representation of a physical object. A wind turbine, for example, is outfitted with several sensors linked to key functional regions. These sensors generate data on the physical thing's performance in various areas, including energy output, temperature, environmental conditions, and more. The information is then transferred and applied to the digital copy by a processing system [19].

5.4   Using Artificial Intelligence to Improve Product Maintenance

Production-related maintenance issues can be challenging. If a machine malfunctions or requires routine maintenance, production must be halted until the problem is resolved. Long downtimes result in lower Return on investment (ROI) and lower customer satisfaction. Because AI systems use sensors to track each machine in a production line, they can detect when a machine begins to act strangely. Many problems could be avoided if the engineers in charge were notified [20].

6   Conclusion

The corporate sustainability and responsibility construct seeks to integrate sustainability and accountability by identifying and addressing the diverse interests of stakeholders. Businesses have the potential to create virtuous loops of beneficial multiplier effects as they grow and succeed. This paper discussed how businesses could pursue economic growth while meeting long-term sustainable goals such as carbon neutrality. We discussed some conceptual frameworks proposed by global initiatives to assess the potential sustainability of business models. Then we shed light on the information technology sector and how businesses in it or closely related to it can reduce their carbon footprint and mitigate the effects of business in general and technology in particular on the climate. Finally, we discussed how businesses could use artificial intelligence to make their businesses more sustainable.

References

1. Rusinko, C.: Green manufacturing: an evaluation of environmentally sustainable manufacturing practices and their impact on competitive outcomes. IEEE Trans. Eng. Manage. 54(3), 445–454 (2007). https://doi.org/10.1109/TEM.2007.900806
2. Morioka, S.N., Evans, S., de Carvalho, M.M.: Sustainable business model innovation: exploring evidences in sustainability reporting. Procedia CIRP **40**, 659–667 (2016)
3. https://unstats.un.org/UNSDWebsite/
4. Parris, T.M., Kates, R.W.: Characterizing and measuring sustainable development. Annu. Rev. Environ. Resour. **28**(1), 559–586 (2003)
5. Brundtland, G.H., et al.: Our Common Future; by World Commission on Environment and Development. Oxford University Press, Oxford (1987)
6. https://epi.yale.edu/
7. https://gsg.org/index.html
8. Emas, R.: The Concept of Sustainable Development: Definition and Defining Principles (2015). https://doi.org/10.13140/RG.2.2.34980.22404
9. Raskin, P., Gallopin, G., Gutman, P., Hammond, A., Swart, R.: Bending the curve: toward global sustainability, Polestar Rep. 8. Stockh. Environ. Inst., Boston, MA. http://www.tellus.org/seib/publications/bendingthecurve.pdf
10. Hare, B.: Fossil Fuels and Climate Protection. The Carbon Logic (1997)
11. Geissdoerfer, M., Vladimirova, D., Evans, S.: Sustainable business model innovation: a review. J. Clean. Prod. **198**, 401–416 (2018)
12. Camilleri, M.A.: Corporate sustainability and responsibility: creating value for business, society and the environment. Asian J. Sustainabil. Social Responsibil. **2**(1), 59–74 (2017). https://doi.org/10.1186/s41180-017-0016-5
13. Quadar, N., Chehri, A., Jeon, G., Ahmad, A.: Smart water distribution system based on IoT networks, a critical review. In: Zimmermann, A., Howlett, R.J., Jain, L.C. (eds.) Human Centred Intelligent Systems. SIST, vol. 189, pp. 293–303. Springer, Singapore (2021). https://doi.org/10.1007/978-981-15-5784-2_24
14. Mikušová, M.: To be or not to be a business responsible for sustainable development? Survey from small Czech businesses. Econ. Res.-Ekonomska Istraživanja **30**(1), 1318–1338 (2017). https://doi.org/10.1080/1331677X.2017.1355257
15. Malmodin, J., Lundén, D.: The energy and carbon footprint of the global ICT and E&M sectors 2010–2015. Sustainability **10**, 3027 (2018). https://doi.org/10.3390/su10093027
16. https://madeintandem.com/blog/environmental-impact-tech-industry/
17. https://www.mapfre.com/en/insights/sustainability/pollution-technology/
18. https://www.ibm.com/case-studies/hera-spa/
19. Botín-Sanabria, D.M., et al.: Digital twin technology challenges and applications: a comprehensive review. Remote Sens. **14**, 1335 (1998). https://doi.org/10.3390/rs14061335
20. Chehri, A., Mouftah, H.: An empirical link-quality analysis for wireless sensor networks. In: 2012 International Conference on Computing, Networking and Communications (ICNC), Maui, HI, USA, pp. 164–169 (2012). https://doi.org/10.1109/ICCNC.2012.6167403
21. Chehri, A., Fortier, P., Aniss, H., Tardif, P.-M.: UWB spatial fading and small-scale characterization in underground mines. In: 23rd Biennial Symposium on Communications, 2006, Kingston, ON, Canada, pp. 213–218 (2006). https://doi.org/10.1109/BSC.2006.1644607
22. Chehri, A., Chaibi, H., Saadane, R., Hakem, N., Wahbi, M.: A Framework of optimizing the deployment of IoT for precision agriculture industry. Procedia Comput. Sci. **176**, 2414–2422 (2020). ISSN 1877–0509, https://doi.org/10.1016/j.procs.2020.09.312
23. Chehri, A., Fortier, P., Tardif, P.: An investigation of UWB-based wireless networks in industrial automation. Int. J. Comput. Sci. Netw. Secur. **8**, 179–188 (2008)

Smart University: Project Management of Information Infrastructure Based on Internet of Things (IoT) Technologies

Yana S. Mitrofanova[1]([✉]), Anna V. Tukshumskaya[2], Svetlana A. Konovalova[3], and Tatiana N. Popova[1]

[1] Togliatti State University, Togliatti, Russia
yana_1979@list.ru
[2] Moscow Pedagogical State University, Moscow, Russia
[3] Moscow State Institute of Culture, Khimki, Russia

Abstract. A smart university (SmU) combines many "smart" and classical educational components based on a single information infrastructure. The basic components of building and developing a smart university are elements of artificial intelligence, the Internet of Things (IoT), intelligent systems tools, neural networks, and other technologies of Industry 4.0. The study examines the issues of improving project management for the development of smart-university components based on IoT elements. IoT elements are basic for the information infrastructure of a smart university. IoT solution development projects have a high level of uncertainty due to the lack of infrastructure development for implementation, an incomplete request from stakeholders. The features of the project management organization are also the dual nature of the IoT product itself, since it has both the software and hardware of the devices. This is not just an IT project based on smart components or SmU information infrastructure. All this requires additional elaboration of methods and tools for managing IoT projects. The article also presents a system of criteria for evaluating the features of IoT projects for SmU and shows their characteristics. The novelty of the research lies developing an adaptive model of a flexible approach to IoT project management. The proposed developments can be applied to managing SmU digital infrastructure and developing projects to create smart components based on IoT. Expert and graphical methods were used for modeling.

Keywords: Internet of Things · IoT · Technologies · Smart University · Smart Components · Project Management · Industry 4.0

1 Introduction and Literature Review

1.1 Managing the Development of IoT Technology as the Main Element of the Digital Smart Infrastructure

The number of IoT devices is constantly growing and will increase to 75.4 billion by 2025 [1]. For the effective functioning of IoT technologies, it is necessary to integrate the work of many elements, including communication systems, sensors, software and

© The Author(s), under exclusive license to Springer Nature Singapore Pte Ltd. 2023
A. Zimmermann et al. (Eds.): KES-HCIS 2023, SIST 359, pp. 101–109, 2023.
https://doi.org/10.1007/978-981-99-3424-9_10

technological platforms that combine devices and allow them to interact with each other. IoT is one of the main infrastructure elements of Industry 4.0 [1]. The development of these technologies causes an increase in the amount of data that needs to be processed using intelligent analysis tools. IoT technologies stimulate the increase of efficiency and productivity of people, give a synergistic effect and change the organizational structures of many systems, including educational ones.

In the field of project management for the creation of SmU components, two approaches are usually considered and proposed for use: classical and flexible [2]. The classical approach to project management has been considered by many modern scientists, such as K. Heldman, R. Newton, and others [3]. Flexible methodologies and features of their implementation are described in the works of K. Schwaber, J. Sutherland, D.A. Loktionov, V.P. Maslovsky, M. Cohn, and others [4].

As a part of the investigation into the hybrid approach, we were based on the research of a number of authors' works that considered various aspects of the application of Agile and its frameworks [5, 6].

The publication of the Project Management Institute (PMI) and the Agile Alliance (Agile Alliance) - "Agile: a Practical guide", which contains recommendations and options for applying the hybrid approach and adapting several methods, can be identified as the most valuable from the point of view of implementing a hybrid approach. This manual is structured in such a way as to correlate with the leading PMI publication, namely the "Guide to the Body of Knowledge on Project Management (PMBoK)" [7].

The instability and uncertainty of the external environment constantly makes changes in project management within the framework of the development of smart universities as an innovation system. Hybrid project management has a set of the most optimal tools for projects implementation for the innovative systems development.

IoT technology and solution development projects have a high level of uncertainty due to the formation of the market, the underdevelopment of the infrastructure for implementation, and an unformed consumers' request. The specifics of project activities organization are also the integrated nature of the IoT product itself: many elements, software and hardware of devices [8, 9]. All this imposes its own peculiarities on the application of methods and tools for project management.

1.2 Analysis of the IoT Ecosystem and Features of IoT Projects for Smart Universities

The interaction between the components of the smart university system is provided by a complex of technologies and solutions from a variety of suppliers included in the ecosystem of IoT technologies [10]. From the point of view of the used technologies, IoT includes the following components:

- Devices and sensors capable of recording events, receiving, analyzing data and transmitting it over the network.
- Means of communication – heterogeneous network infrastructure combining heterogeneous communication channels – mobile, satellite, wireless, and fixed.
- IoT platforms designed to manage devices and communications, applications and analytics.

– Applications and analytical software responsible for data processing, creating predictive models, and intelligent device management.
– Data storage systems and servers capable of aggregating, storing and processing large amounts of various information.
– Services for the development or adaptation of IoT solutions that require knowledge of the industry and the specifics of the customer's business.
– Security solutions responsible for the security of the entire operational process.

A generalized model of the IoT ecosystem is presented in Fig. 1.

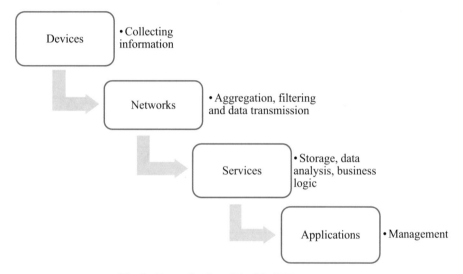

Fig. 1. Generalized model of the IoT ecosystem.

Based on the conducted research, the factors and interrelations determining the features of IoT projects were systematized. The developed system is shown in Fig. 2.

The listed features of the projects impose their own specifics on the choice of methodology and tools for project management, the construction of business models and building of a smart university development strategy.

1.3 The Problem, Goals and Objectives of the Study

The problem of the research is to find mechanisms and approaches for building an effective information infrastructure of a smart university based on IoT technologies. IoT solution development projects have a high level of uncertainty due to the lack of infrastructure development for implementation, and an unformed request from stakeholders. This requires additional elaboration of IoT project management methods and tools.

The article also presents a system of criteria for evaluating the features of IoT projects for SmU and shows their characteristics. The novelty of the research lies in the development of an adaptive model of a flexible approach to IoT project management.

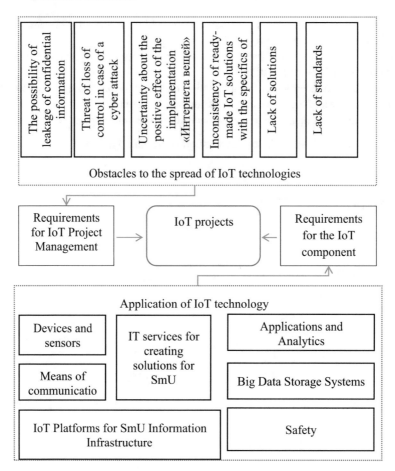

Fig. 2. Features of IoT projects for SmU

The proposed developments can be applied to the management of SmU digital infrastructure and the development of projects to create a smart components based on IoT.

The main purpose of this study is to obtain a system of criteria for evaluating the features of IoT projects for SmU and to develop an adaptive model of a flexible approach to IoT project management aimed at removing uncertainty and improving the efficiency of the development of the information infrastructure of smart universities.

The objectives of the study are:

1. Analysis of IoT technology as the main element of the smart digital infrastructure.
2. Analysis of the IoT ecosystem and features of IoT projects for smart universities.
3. Modeling the system of criteria for evaluating the features of IoT projects for SmU.
4. Development of an adaptive model of a flexible approach to IoT project management.

2 The Concept of Forming a Criteria System for Evaluating the Features of IoT Projects and Their Implementation for the Information Infrastructure Development of a Smart University

Taking into consideration the studied features of IoT projects, it is necessary to form a system of recommendations for the use of hybrid models using the best of two popular and often opposite approaches (classical and flexible) to manage the SmU information infrastructure and develop projects for creating smart components based on IoT.

A system of criteria for evaluating IoT projects is proposed. It is based on the evaluation system of D.A. Loktionov [11], taking into account the specifics of IoT projects (Fig. 3).

The criteria system for evaluating IoT projects is presented as a graphical model. In the future can also be used for a survey to determine the IoT project flexibility with the possibility of digitizing responses and applying Agile elements.

The description of the criteria system is presented in Table 1.

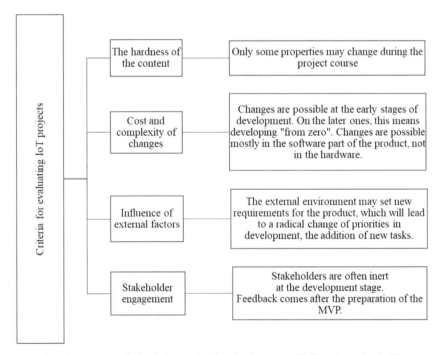

Fig. 3. System of criteria for evaluating the features of IoT projects for SmU

Thus, the survey and the results interpretation of the expert assessment based on the table will allow to determine the applicability of Agile methodology in IoT projects for the smart university information infrastructure, to form a criteria system for evaluating the features of IoT projects, as well as to identify "pain points" in project management

Table 1. Description of IoT project evaluation criteria for SmU

Evaluation criteria	Features of the Internet of Things projects
"The hardness of the content"	The goal and product of the project are clear at the start. During the project, only some properties of the product (SmU components) may change, the inclusion of which, in terms of reference, does not significantly affect the structure of work and resources spent on development. Otherwise, the development may go from cycle to cycle and be greatly delayed, leading to the closure of the project due to a lack of results
"Cost and complexity of changes"	Changes are possible at certain stages of development. In particular, at the stage of the very first versions of prototypes when determining the basic modification of the SmU component. At later stages, making changes may mean developing from scratch At the moment, the development of models with the possibility of modification in the future is being practiced. However, this also has its limitations. In particular? these modifications are possible mostly in the software part of the product and only to a small extent in the hardware
"The influence of external factors"	The external environment can quickly and rigidly set new requirements for the product. This can lead to a radical change of priorities in development, the addition of new tasks, and accordingly, delay the development of components for SmU
"Stakeholder engagement"	Many stakeholders are inert at the development stage and do not express their requirements for the product at the start of the project. Feedback is received only after the preparation of the MVP, which carries the risk of the product failing to meet the expectations of stakeholders

at the moment. Based on the analysis, it can be recommended to continue to adhere to the classical approach in IoT project management using Agile elements, introducing a hybrid approach.

3 Development of an Adaptive Flexible Approach Model to IoT Project Management

Considering the concept of forming a system of criteria for evaluating the IoT projects features, an adaptive model of a flexible approach to managing these projects was developed. This model can be used to develop the information infrastructure of a smart university.

The model is designed for IoT projects. Its main difference from the existing flexible approach models to project management is that the model takes into account the division of product development into two areas: software and hardware development. The hardware part is based on a predictive, classical approach, and the software part is flexible. The developed model is universal and can be used by other organizations where IoT project management is being implemented.

The transition to a hybrid lifecycle will allow the team to plan the delivery of value taking into account the existing risks. The delivery of value will happen earlier, and it will allow us to get feedback faster, work out new requirements and prepare the product and eventually move to the commercialization phase. At the same time, resources are saved, and the return on investment is accelerated (Fig. 4).

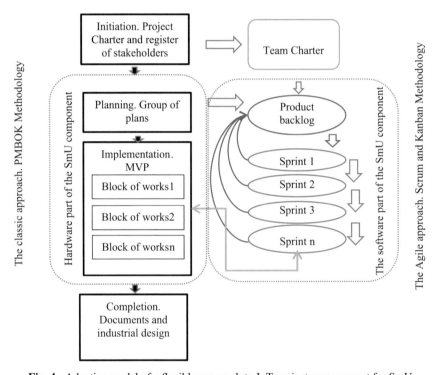

Fig. 4. Adaptive model of a flexible approach to IoT project management for SmU

Many teams are not able to switch to ways of working on Agile principles in one day. The gradual transition is associated with the addition of iterative methods to improve knowledge sharing and consistency between teams and stakeholders. In the future, the model can be developed by including incremental methods in order to accelerate the delivery of value and return on investment.

4   Conclusion and Next Steps

Conclusions

1. Using a hybrid approach in IoT project management makes it possible to offset the shortcomings of the classical methodology in IoT project management for building an effective information infrastructure of a smart university.
2. In this study, the task of highlighting the distinctive features of IoT technology projects was achieved. They include the multi-component of the final product of IoT technology and, as a consequence, the variety of roles of market participants; the strict dependence of the developed product on the infrastructure presented on the market; standards inconsistency in the field of IoT, doubts about the effectiveness of IoT solutions on the part of the consumer, etc.
3. An adaptive model of a flexible approach to IoT project management has been developed. This model can be used to develop the information infrastructure of a smart organization. Its main difference from the existing models of a flexible approach to project management is that the model takes into account the division of product development into two areas: software and hardware development.

Next Steps

1. Further study and development of IoT project management tools and methods on the use of various combinations and adaptations of "flexible" tools in project activities within the framework of developing information infrastructure of smart organizations.
2. Working on integration issues of IoT components within the framework of information infrastructure platforms of smart organizations.

References

1. Schwab, K., & Davis, N. Shaping the future of the fourth industrial revolution. Currency. (2018)
2. Morcov, S., Pintelon, L., Kusters, R.: Definitions, characteristics and measures of IT project complexity - a systematic literature review. Int. J. Inf. Syst. Proj. Manag. **8**(2), 5–21 (2020)
3. Mitrofanova, Y.S., Burenina, V.I., Tukshumskaya, A.V., Popova, T.N.: Project Management as a Tool for Smart University Creation and Development. In: Uskov, V.L., Howlett, R.J., Jain, L.C. (eds.) Smart Education and e-Learning 2020. SIST, vol. 188, pp. 317–326. Springer, Singapore (2020). https://doi.org/10.1007/978-981-15-5584-8_27
4. Mitrofanova, Y.S., Chehri, A., Tukshumskaya, A.V., Vereshchak, S.B., Popova, T.N.: Project Management of Smart University Development: Models and Tools // Smart Innovation. Systems and Technologies **240**, 339–350 (2021)
5. Smart University: Digital Development Projects Based on Big Data / Y. S. Mitrofanova, V. I. Burenina, A. V. Tukshumskaya [et al.] // Smart Innovation, Systems and Technologies. –Vol. 305 SIST. – P. 230–240. – DOI https://doi.org/10.1007/978-981-19-3112-3_21 (2022)

6. Tam, C., Moura, E. J. d. C., Oliveira, T. and Varajão, J. The factors influencing the success of on-going agile software development projects. International Journal of Project Management, 38, 3, 165–176 (2020)

7. Takagi, Nilton & Varajão, João. Success Management and the Project Management Body of Knowledge (PMBOK): An Integrated Perspective // https://www.researchgate.net/public ation/347949303_Success_Management_and_the_Project_Management_Body_of_Know ledge_PMBOK_An_Integrated_Perspective_-research-in-progress (2020)

8. Alsharif, Mohammed H.; Kelechi, Anabi H.; Yahya, Khalid; Chaudhry, Shehzad A. Machine Learning Algorithms for Smart Data Analysis in Internet of Things Environment: Taxonomies and Research Trends // Symmetry 12, no. 1: 88. (https://doi.org/10.3390/sym12010088) (2020)

9. Chehri, A. and Mouftah, H. An empirical link-quality analysis for wireless sensor networks, 2012 International Conference on Computing, Networking and Communications (ICNC), Maui, HI, USA, pp. 164–169, doi: https://doi.org/10.1109/ICCNC.2012.6167403 (2012)

10. Chehri, A., Jeon, G.: The Industrial Internet of Things: Examining How the IIoT Will Improve the Predictive Maintenance. In: Chen, Y.-W., Zimmermann, A., Howlett, R.J., Jain, L.C. (eds.) Innovation in Medicine and Healthcare Systems, and Multimedia. SIST, vol. 145, pp. 517–527. Springer, Singapore (2019). https://doi.org/10.1007/978-981-13-8566-7_47

11. Loktionov, D.A., Maslovskiy, V.P. Criteria for applying the Agile methodology for project management. Creative economy12(6), 839–854. doi: https://doi.org/10.18334/ce.12.6.39179 (2018)

Smart Manufacturing: Intelligent Infrastructure Based on Industry 4.0 Technologies

Yana S. Mitrofanova[1]([✉]), Valentina I. Burenina[2], Vladimir G. Chernyh[3], and Tatiana N. Popova[1]

[1] Togliatti State University, Togliatti, Russia
yana_1979@list.ru
[2] Bauman Moscow State Technical University, Moscow, Russia
[3] Federal State Budget Educational Institution of Higher Professional Education, «Platov South-Russian State Polytechnic University (NPI)», Novocherkassk, Russia

Abstract. In contrast to existing literature, this study examines a smart production management system based on bid data and other Industry 4.0 technologies. Smart components of smart production continuously generate a very large amount of data, and all this data must considered when making management decisions, efficiently processed, and stored. At the same time, the article does not consider the essence of the structure of smart production management systems. It discusses an intelligent control system. The article also discusses the modern production experience of digital transformation. It also offers solutions for creating an efficient digital production infrastructure and decision support. Attention is focused on the information infrastructure of smart manufacturing, which allows you to take advantage of smart technologies such as big data, machine learning, and the industrial Internet of Things (IoT). At a base of the infrastructure for storing and processing information and knowledge, it is recommended to allocate a component for storing big data. As such a component, it would be most optimal to use a data lake. BPMS technology can become the core of the smart manufacturing information support infrastructure and the center of information technology integration.

Keywords: Internet of Things · Technologies · Digital Transformation of Industry · Digital Infrastructure · Level of Digital Maturity · Artificial Intelligence · RPA · IoT · Big Data · BPMS · Industry 4.0

1 Introduction and Literature Review

1.1 Analysis of the Pandemic Problems and Consequences for the Industry and Its Digitalization Processes

The digital economy and the development of industry 4.0 ideas have an impact on all sectors of the economy, including industry. Currently, the most competitive will be those industrial enterprises that can quickly and efficiently go through the stages of digital transformation and take advantage of IoT technologies, big data, artificial intelligence, digital twins, the industrial Internet of things, and others.

© The Author(s), under exclusive license to Springer Nature Singapore Pte Ltd. 2023
A. Zimmermann et al. (Eds.): KES-HCIS 2023, SIST 359, pp. 110–119, 2023.
https://doi.org/10.1007/978-981-99-3424-9_11

At the same time, it is necessary to understand that the industry is fairly conservative, and the introduction of any innovations, including digital ones, is quite slow and gradual. Many enterprises are limited in financial resources; others refuse technological transformations. The pandemic also affected the state of enterprises and highlighted the problems of digital immaturity in industrial systems.

According to the research [1, 2], the main problem of industry digital transformation is the high cost of digital projects, as well as the initial low level of digitalization of production systems and resistance to changes on the part of personnel. The survey participants noted that the COVID pandemic and the digital solutions found during it, help to solve the issues of improving enterprise management using digital tools, and that the digitalization of production was practically unaffected. Also, in the study, according to the survey data, the main directions for solving the problems listed above are highlighted (Fig. 1).

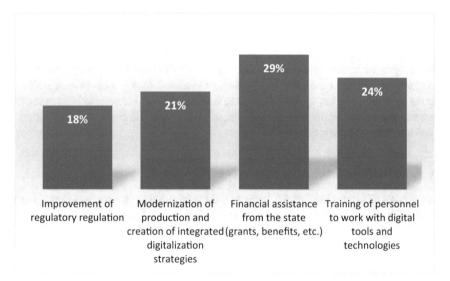

Fig. 1. The main directions for solving the problems of digital transformation in industry (according to a survey of 251 respondents from the expert community of industry in Russia and India) (%) [1]

The greatest interest among manufacturers in the process of digital transformation is caused by production technologies, including technologies for resource management, organizational capabilities, life cycle management, and production product data. Secondly, they are interested in process robotics technologies (RPA), and thirdly, Industrial Internet of Things (IIoT) technologies [3], artificial intelligence (AI), and others.

1.2   Digital Analytics and the Introduction of a Cybernetic Approach to the Industrial Enterprise Management

Nowadays, the process of accumulating experience, mastering the capabilities, and understanding the prospects of new technologies, including big data technologies, continues. Digital data analytics allows a manufacturing enterprise to make management decisions based on big data in real-time.

The Global Big Data Analytics Market (BDA) for 2020 believes that the demand for big data analytics will grow exponentially. Data security is a major concern in all industries due to the growing deployment of the Internet of Things (IoT) and the increasing number of devices that create huge amounts of data.

The worldwide BDA market is estimated to grow 4.5 times, collecting revenues of $68.09 billion by 2025 from $14.85 billion in 2019, with an average annual growth rate of 28.9% [4]. In addition, amid the uncertainty associated with the pandemic, BDA continues to be one of the top deployment priorities for many enterprises, as its use will help them remain competitive while accelerating innovation, including in industry.

In accordance with the ideas of Industry 4.0, the introduction of a cybernetic approach to management, which is based on decision-making based on the results of objective data analysis (data-driven decision), will get rid of the "disease" of any HiPPO-type management systems (Highest-Paid Person's Opinions) [5, 6]. This rule of decision-making is inherent in any system, including industrial ones, where officials (managers) often make far from optimal decisions. These management decisions are not based on up-to-date data and analytics.

1.3   The Problem, Goals and Objectives of the Study

The question of improving the efficiency of the management system arises in the context of limited financial resources and the need for digital transformation of manufacturing enterprises. It is necessary to assess the enterprise's digital maturity and redistribute available resources in order to accelerate digital transformation processes and implement the most effective projects in a timely manner. Management support in the process of digital transformation and the development of smart production should be an analytical system.

The analytical system of digital product management is primarily based on data. Smart components of a manufacturing enterprise based on Industry 4.0 technologies generate a huge amount of data. At the same time, any data is collected and stored. In a conventional production management system, data analytics is performed mainly on demand, which does not take into account the huge potential inherent in big data. With the right approach to the use of big data, they can become the basis of a smart analytics system for managing a production system.

Therefore, the research problem can be described as follows: how to determine the level of digital maturity and design an optimal intelligent digital transformation project management system based on big data to continue supporting and promoting digitalization processes and increasing the intelligence of the production system.

Our research will focus on the development of infrastructure and an intelligent management system for a manufacturing enterprise based on big data.

2   Levels of Digital Maturity of an Industrial Enterprise

Digital transformation is a catalyst for the formation and achievement of the goals of transitioning from one state of development of an industrial enterprise to another, more qualitative one. The digital transformation completes the transition from the classical form of production to smart production.

There are different approaches to determining the level of digitalization in an industrial enterprise. The models of digital maturity developed by Booz and Company are interesting (three levels of digital maturity are distinguished), Lichtblau et al. (3 levels of digital maturity are allocated), McKinsey & Company (5 levels are allocated), Forester (4 types are allocated according to the digital maturity index), and others [7, 8].

Based on existing research [8, 9] and open data of industrial enterprises, in our opinion, the following levels of digital maturity of an industrial enterprise can be distinguished during the transition to smart production: initial digitalization, digital manageability, and digital transformation. At the same time, it should be noted that in the process of transitioning from one level of maturity to another, the objects of transformation are not only the business processes of management and production of the enterprise but also data, system interfaces, and personnel (Fig. 2).

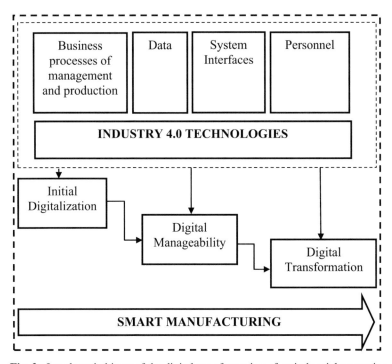

Fig. 2. Levels and objects of the digital transformation of an industrial enterprise

At the stage of initial digitalization, information systems have already been introduced at the enterprise that automate the main and auxiliary processes (management and production processes). Digital accounting should be introduced, data digitized, work with digital data organized, and their generation organized. Users of information systems and portals (enterprise personnel) at the stage of initial digitalization should have the basic skills to work on a computer and in specific information systems (CRM, ERP, MES, RFID, and others) installed at an industrial enterprise [10, 11].

The stage of digital manageability is characterized by the presence of a single information base, where all the data of a manufacturing enterprise is collected, a consistent data info model is implemented, integrity rules apply, a digital footprint is fixed (API and other tools), industrial robots are partially implemented, information systems are integrated (CRM, ERP, MES, EMI, NSI systems, design and technological systems, etc.), automatic execution of processes with the formation of KPIs is organized, and paperless document flow with the use of an electronic digital signature is organized. Personal cabinets and desktops with notifications and tracking of all actions have been created for all participants in the production and the management processes. At the stage of digital manageability, all digital tools have mobile and ergonomic interfaces. Users of digital tools at the stage of digital manageability should have the skills for the digital transformation of information, work with knowledge management tools, use analytics tools (for example, EMI), and be able to build digital communications.

The next level of digital maturity (digital transformation) is characterized by the introduction of intelligent digital tools for working with big data; BDA and EMI tools are used; robotization of processes is provided (industrial robots, expert systems, and artificial intelligence tools are used; chatbots and RPA elements are introduced); all interfaces are integrated and intelligent services are provided. Digital profiles of personnel and management have been formed. A digital profile management toolkit has been implemented and is working. In our opinion, these are one of the main signs of the digital transformation of a manufacturing enterprise in the transition to smart manufacturing.

Based on these characteristics of the levels of digital maturity, a number of industrial enterprises were surveyed based on open-source data and a survey of experts for the presence of certain signs of digital transformation and a strategy for the development of smart manufacturing. The study's findings are depicted in Fig. 3.

A self-examination of an industrial enterprise and expert assessments can be used as an additional way to determine the level of digital maturity. The further development of this research will be the formation of mathematical models to improve the quality of assessment and the use of big data technologies [8].

To reach these levels of digital maturity, it is necessary to use advanced management technologies [12] that will accelerate the decision-making process, help choose the right strategy for the transition to smart manufacturing, and develop smart manufacturing infrastructure.

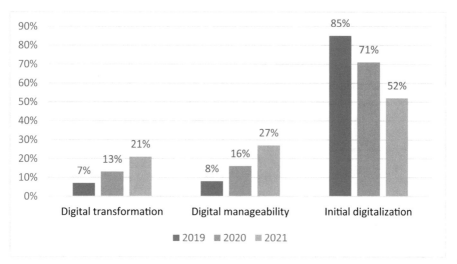

Fig. 3. Determination of the level of digital maturity of the studied enterprises in dynamics for 2019–2021

3 Designing the Digital Infrastructure of Smart Manufacturing

The high level of intellectualization and robotization of production, the studio of digital manageability, and the studio of digital transformation are characterized by the introduction of smart analytics, the intellectualization of basic and auxiliary business processes, the integration of all systems and interfaces, and more [13, 14]. And all these components of the digitalization of an industrial enterprise should be integrated on the basis of a single infrastructure. This infrastructure should work with big data, forming the analytical foundation of an intelligent enterprise management system and providing a quick and accurate real-time assessment based on a large amount of data and their analysis [15].

The generalized structure of the intelligent control system is shown in Fig. 4.

The infrastructure for storing and processing information and knowledge should be based on a component for storing big data. As such a component, it would be most optimal to use a Data Lake [16]. The big data of an industrial enterprise includes a different format for presenting information: MES, CRM, ERP, APIs data, regulatory reference information (NSI), sensor and sensor data, multimedia files, records from databases, and more. To extract useful information for the management system from all this data, it must first be collected. A data lake is suitable for this. It is a repository for a large volume of unstructured data collected or generated by a production system. Unlike corporate data warehouses or data warehouses, unstructured raw data is stored in the data lake. Such data can be used for rapid decision-making and event prediction using machine learning algorithms.

Smart enterprise services allow you to collect a lot of data based on digital counterparts, personnel, management, and other stakeholders of the enterprise. The data lake accumulates all digital activities related to production processes, personnel, and other business processes and smart components of the enterprise information system.

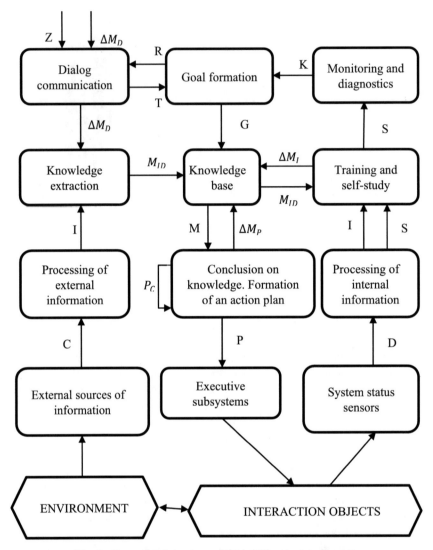

Fig. 4. Generalized structure of the intelligent control system

BPMS (Business Process Management System) technology, the structure of which is shown in Fig. 5 [15, 16], can become the core of the information support infrastructure for smart manufacturing and the center for the integration of information technologies.

In order to develop the infrastructure and analytical management subsystem, including within industrial enterprises, the ISO/IEC TR 20547–1:2020 Information Technology—Big Data Reference Architecture—Part 1: Framework and Application Process standard has been developed. The document contains a description of the structure of the reference architecture of the system for working with big data and also provides a solution to the problem of displaying possible big data use cases in the reference

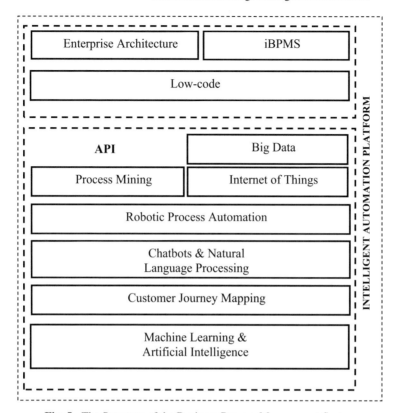

Fig. 5. The Structure of the Business Process Management System

architecture. The provisions of the standard can be applied in enterprises to describe the architecture of specific systems for working with big data and the implementation of these systems, taking into account the technologies used as well as the roles and needs of the performers.

4   Conclusion and Next Steps

Conclusions

As part of the study, the following results were obtained:

1. A model for assessing the digital maturity of an industrial enterprise on the way to the transition to the concept of smart production is proposed, which will correctly determine the strategy for further digitalization and redistribute financial resources to the most effective digital transformation projects.
2. The development directions of the smart production management system based on big data have been determined.

3. The concept of digital infrastructure formation through smart production development based on an intelligent system is proposed.

Next Steps

1. Research of IoT technologies and the possibility of their integration to improve the efficiency of the digital infrastructure of smart manufacturing.
2. The further development of the assessment of the digital maturity of an industrial enterprise will be the formation of mathematical models as a way to improve the quality of the assessment and the study of the possibilities of using big data technologies in this direction.

References

1. Digitalization industry 2022. TAdviser (2022). https://www.tadviser.ru/index.php/
2. Chehri, A., Popova, T.N., Vinogradova, N.V., Burenina, V.I.: Use of innovation and emerging technologies to address Covid-19-like pandemics challenges in education systems. In: Uskov, V.L., Howlett, R.J., Jain, L.C. (eds.) Smart Education and e-Learning 2021. SIST, vol. 240, pp. 441–450. Springer, Singapore (2021). https://doi.org/10.1007/978-981-16-2834-4_38
3. Chehri, A., Jeon, G.: The industrial Internet of Things: examining how the IIoT will improve the predictive maintenance. In: Chen, Y.-W., Zimmermann, A., Howlett, R.J., Jain, L.C. (eds.) Innovation in Medicine and Healthcare Systems, and Multimedia. SIST, vol. 145, pp. 517–527. Springer, Singapore (2019). https://doi.org/10.1007/978-981-13-8566-7_47
4. Global Big Data Analytics Market to Grow 4.5 Times by 2025. Powered by Data Security Requirements. https://ww2.frost.com/news/press-releases/global-big-data-analytics-market-to-grow-4-5-times-by-2025-powered-by-data-security-requirements/
5. Schumacher, A., Erol, S., Sihn, W.: A maturity model for assessing Industry 4.0 readiness and maturity of manufacturing enterprises. Procedia CIRP **52**, 161–166 (2016). https://doi.org/10.1016/j.procir.2016.07.040
6. Oesterreich, T.D., Teuteberg, F.: Understanding the implications of digitisation and automation in the context of Industry 4.0: a triangulation approach and elements of a research agenda for the construction industry. Comput. Ind. **83**, 121–139 (2016)
7. Remane, G., Hanelt, A., Wiesboeck, F.: Digital maturity in traditional industries – an exploratory analysis. In: Twenty-Fifth European Conference on Information Systems (ECIS), Guimarães, Portugal (2017). https://www.researchgate.net/publication/316687803
8. Mitrofanova, Y.S., Burenina, V.I., Tukshumskaya, A.V., Kuznetsov, A.K., Popova, T.N.: Smart University: digital development projects based on big data. In: Uskov, V.L., Howlett, R.J., Jain, L.C. (eds.) Smart Education and e-Learning - Smart Pedagogy, pp. 230–240. Springer Nature Singapore, Singapore (2022). https://doi.org/10.1007/978-981-19-3112-3_21
9. Mitrofanova, Y.S., Glukhova, L.V., Burenina, V.I., Evstafeva, O.A., Popova, T.N.: Smart production: features of assessing the level of personnel digital readiness. Procedia Comput Sci **192**, 2962–2970 (2021)
10. Chehri, A., Mouftah, H.: An empirical link-quality analysis for wireless sensor networks. In: 2012 International Conference on Computing, Networking and Communications (ICNC), Maui, HI, USA, pp. 164–169 (2012). https://doi.org/10.1109/ICCNC.2012.6167403

11. Shemshura, E.A.: Development of modelling algorithm of technological systems by statistical tests. In: Shemshura, E.A., Otrokov, A.V., Chernyh, V.G. (edn.). IOP Conference Series: Materials Science and Engineering : Processing Equipment, Mechanical Engineering Processes and Metals Treatment, Tomsk, 04–06 декабря 2017 года. vol. 327, 4. – Tomsk: Institute of Physics Publishing, p. 042072 (2018). https://doi.org/10.1088/1757-899X/327/4/042072

12. Mitrofanova, Y.S., Chehri, A., Tukshumskaya, A.V., Vereshchak, S.B., Popova, T.N.: Project management of smart university development: models and tools. Smart Innovation Syst. Technol. **240**, 339–350 (2021)

13. Sergeeva, M.G., Bedenko, N.N., Tsibizova, T.Y., Mohammad Anwar, M.S., Stanchuliak, T.G.: Organisational economic mechanism of managing the growth of higher education services quality. Espacios **39**(21) (2018)

14. Sergeeva, M.G., Bedenko, N.N., Karavanova, L.Z., Tsibizova, T.Y., Samokhin, I.S., Mohammad Anwar, M.S.: Educational company (technology): Peculiarities of its implementation in the system of professional education. Espacios **39**(2) (2018)

15. Alsharif, M.H., Kelechi, A.H., Yahya, K., Chaudhry, S.A.: Machine learning algorithms for smart data analysis in Internet of Things environment: taxonomies and research trends. Symmetry **12**(1), 88 (2020). https://doi.org/10.3390/sym12010088

16. Llave, M.R.: Data lakes in business intelligence: reporting from the trenches. Procedia Comput. Sci. **138**, 516–524 (2018)

Towards an Optical IoT-Based Power Transformer's Insulating Paper Monitoring

N. Seifaddini[1]([envelope]), K. S. Lim[2], O. C. Weng[2], W. Udos[2], B. Sekongo[1], U. Mohan Rao[1], F. Meghnefi[1], I. Fofana[1], and M. Ouhrouche[1]

[1] Research Chair on the Aging of Power Network Infrastructure (ViAHT), University of Quebec at Chicoutimi, Chicoutimi, QC G7H 2B1, Canada
najmeh.seifaddini1@uqac.ca
[2] Photonics Research Centre, University of Malaya, Kuala Lumpur, Malaysia

Abstract. The insulation system of a transformer consists of oil-paper material that significantly affects its lifespan. Even under normal operations, the oil-paper insulation inevitably deteriorates over time. Measuring the degree of polymerization (DP) is a common and direct method to evaluate the degradation of the paper insulation. However, it can be challenging to access this information once the transformer is in use. Sampling the insulating oil for analysis can provide information but sampling the solid insulation system (paper) is not always possible. To address this issue, it is important to monitor the solid insulation system indirectly. Furthermore, the lifespan of a transformer is tied to the condition of its paper insulation as it ages. In this study, an optical aging marker is explored as an alternative to monitor the degradation of the paper insulation in oil-filled transformers. The reflectance of aged papers was monitored, and the relationship with DP was investigated at three different wavelengths. The results showed the reflectance spectrum and DP values have a correlation that can be used to monitor the aging of the paper insulation.

Keywords: Solid insulation · Degree of polymerization · Reflectance

1 Introduction

Transformers play a crucial role in power generation facilities, distribution networks, and transmission systems [1, 2]. Power transformers typically run uninterrupted for many years, with maintenance personnel relying mostly on the physical safeguards of the transformer and carrying out limited upkeep using conventional methods. However, due to factors such as increased load demands, power quality requirements, environmental regulations, and equipment maintenance goals set by management; there is now a need for comprehensive monitoring and assessment systems that are available online. Replacing transformers is a costly and time-consuming process, so any strategies that can extend their lifespan would be considered a wise investment in the long term [3, 4].

The transformer insulation system is typically made up of cellulose paper and oil materials, which are subjected to various stresses, including thermal, electrical, mechanical, and chemical impacts, during operation. The safety of the transformer relies heavily

© The Author(s), under exclusive license to Springer Nature Singapore Pte Ltd. 2023
A. Zimmermann et al. (Eds.): KES-HCIS 2023, SIST 359, pp. 120–129, 2023.
https://doi.org/10.1007/978-981-99-3424-9_12

on the insulation components being in good condition, which is why reliable evaluation tools to assess the insulation system during operation are crucial. The state of the solid insulation is widely recognized as an indicator of the overall condition of the power transformer [5–7].

The breakdown of paper insulation results in various decomposition by-products, including water, alcohols, carbon oxides, furanic derivatives, and other organic compounds. The presence of these dissolved compounds in the oil can be analyzed and used as indicators or chemical markers to assess the degradation of cellulosic paper insulation [8]. These chemical markers can be utilized as tools for continuously monitoring the condition of the solid insulation in the transformer, providing insights into its lifespan [8, 9].

It is important to note that evaluating the condition of solid insulation can be challenging because it is not easily accessible in operating power transformers. Researchers are exploring indirect methods, such as analyzing chemical markers of cellulose degradation, to address this issue [10]. However, a major drawback of these tracers is the loss of information once the transformer oil is regenerated/reclaimed [11]. Another issue is the increase in local heating and surface tracking under operating electrical stress. Currently, there is no practical method for direct measurement of moisture in the paper. Although, in theory, measuring the water content of the transformer oil can give an estimate of the moisture content in the paper when in equilibrium, this state is never achieved in an operating transformer [12].

The cellulose paper used in transformer insulation is made up of long chains of glucose rings, known as cellulose polymer molecules. The average length of these chains is referred to as the degree of polymerization (DP). While the DP can be measured directly from a paper sample, this is not practical for an operating power transformer, as it would require an invasive manipulation of the unit. As an alternative, chemical methods can be used to estimate the DP value indirectly [8, 13].

The DP is a direct indicator of the correlation between degradation of the insulation paper and its mechanical strength. To assess the quality of the cellulose, viscometric methods are employed that relate intrinsic viscosity to molecular weight. The paper changes colour to dark brown in the range of 200 to 250, and in the range of 150 to 200, it loses its mechanical strength completely. Eventually, electrical breakdown and transformer failure can occur [14, 15].

The insulation system can be monitored in accordance with International Electrotechnical Commission (IEC) and American Society for Testing and Materials (ASTM) standards. Some of the tests and measurements that comply with these standards include dissolved gas analysis (DGA), detection of furanic compounds, methanol content, and depolymerization value (DP) [16, 17]. However, these precise diagnostic methods are both time-consuming and costly. As a result, alternative methods for monitoring the insulation systems of power transformers have been proposed. Optical methods, which are non-destructive and hold great potential, have garnered the attention of researchers and have been demonstrated as an alternative to conventional methods [17].

The intricacy involved in the design, construction, operation, and environmental factors of power transformers makes it challenging to assess their conditions. Fibre-optic sensors have emerged as a popular and growing method for monitoring the insulation paper and oil in transformers [18, 19]. Optical sensors, capable of integrating with technological advancement, are gaining interest due to a consequent reduction in design costs, immunity to stray magnetic/electric fields, and improved technical portability [11]. Using sensor technology, it is possible to reliably determine the depolymerization value. An optical system based on the principles of optical spectroscopy directs the light reflected from the paper into the spectrometer. An optical system based on optical spectroscopy principles reflects the light from the paper into a spectrometer. An online monitoring system that assesses the insulation conditions of the transformer paper offers numerous benefits [20].

The Internet of Things (IoT) technology was developed after the computer, internet, and mobile communications to bring intelligence to identification, location, tracking, monitoring, and management in a network. Currently, the advancement of IoT technology and sensors, as well as their combination, are being considered. Fiber optic (FO) sensors are one of the best sensing techniques due to their exceptional and specific features. Optical fiber-based online systems consist of users, the Internet, a cloud platform, and interconnected FO sensors that are responsible for sensing. Data such as temperature, moisture, vibration, etc. can be collected through the FO sensors, uploaded to the control room via the transmit protocol, and analyzed before being transmitted to the users [21]. The combination of optical fiber sensors and the Internet of Things (IoT) can create an accurate online monitoring system, well suited for smart grid applications [22, 23].

This study evaluates the feasibility of using an optical aging marker to assess the quality of paper insulation in oil-filled transformers. The research involved aging paper samples at various intervals in a mechanical convection oven, and analyzing the relationship between the optical reflectance spectrum and the extent of polymerization of the paper insulation. The goal was to determine if changes in the optical reflectance spectrum could be used as an indicator of the aging process of the paper insulation.

2 Experimental

2.1 Thermal Aging

In the present study, electrical grade cellulose kraft-type is used, with thicknesses of 0.18 mm and 0.25 mm. Initially, 980 and 963 degrees of polymerization were achieved for 0.18 mm and 0.25 mm thicknesses, respectively. Twin-blade cutters (TMI brand) were used to calendar Kraft papers. Dehydration of the strips took place at 105 °C under a vacuum for 48 h. In the following step, natural ester (1204) and synthetic ester (Midel 7131) oils were used to impregnate paper samples for 24 h at room temperature. An accelerated thermal aging process was conducted with aging ampoules prepared with a 10:1 oil-paper mass ratio and placed in a mechanical convection oven at 150 °C.

Thermal aging experiments with an open beaker were performed after three weeks, four weeks, five weeks, and six weeks with a controlled aging history. An individual paper strip is homogeneously aged if its aging is uniform. Copper spacers were used to ensure uniform contact between the paper strips and the insulating oil. To minimize oxygen and moisture entry into the containers (aging ampoules), stainless-steel lids were used. It would be better to mention that before starting tests the aged samples were degreased.

2.2 Degree of Polymerization (DP)

The DP values were specified in the laboratory according to ASTM D-4243. The DP values were obtained for two different thicknesses of paper samples aged under controlled accelerated thermal aging in two different oils: 1204 (NE) and 7131 (SE). It is to be mentioned that each value represents an average of three measurements. Figure 1 shows the DP values of paper samples were significantly decreased due to thermal stress at 150 °C. The result of DP measurements that the paper of 0.25 mm in NE oil has a higher DP, which is approximately 500 for the most aged. Therefore, papers aged in SE oil under the described conditions have a higher DP and better conditions.

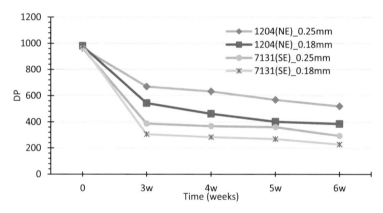

Fig. 1. Degree of polymerization (DP) of the Kraft papers as a function of thermal aging.

2.3 The Reflectance Measurement Setup

In this study, a high-performance L-module multifunction diode with strong output power, remarkable stability, and a broad selection of wavelength options was utilized. The setup also employed a standard SMA connector, along with a 300/330 m large-core glass cable and a 3.0 mm outer diameter jacket. Fibre-optic collimators are tools that direct light from an optical fiber into a collimated beam in open space or collimated light into a fiber. A Thorlabs S140C photodiode power meter sensor, which acts as a light detector, was used to determine the optical spectrum. The thickness, refractive index, and type of oil used for impregnation and aging, all play a role in determining the reflectance spectrum of the insulation paper. Figure 2 provides a schematic illustration of the laboratory scale utilized for the reflectance measurement setup.

The incident beam was emitted from the light source used to illuminate the insulation paper through the fibre-optic cable. For analytical purposes, the reflected ray is collected and sent to an optical spectrometer.

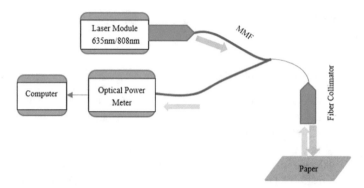

Fig. 2. The experimental setup used to measure the reflectance of paper samples.

As seen in Fig. 2, the first collimator allowed the source's light to be transmitted into an MMF (MultiMode Fiber). The second collimator directs the collimated light beam to the paper surface. A portion of the entire spectrum is reflected, collected, and focused by the second collimator onto an MMF, which is then picked up by the optical power meter. To ascertain the insulation paper's optical characteristics, the reflected ray is examined. The DP and reflectance of the papers under various conditions were correlated appropriately.

3   Results and Discussion

Reflectance from the paper surface was measured at different aging levels. Figure 3 shows the reflectance spectrum response of both paper thicknesses under various aging conditions in ester fluids.

The number of hydroxyl and ether groups decreased during the degradation. The oxidized hydroxyl groups form carbonyl and carboxyl groups. In other words, the hydrogen bonds between the cellulose molecules and the hydroxyl groups within the cellulose molecules are broken [24]. The scission of the monomer chains reduces the paper's DP value and thus its mechanical and optical qualities. According to the description above, due to the fragility of the paper's lattice and decreasing mechanical and physical strength of the paper, light easily passes through the aged and weakened paper and the reflectance of the paper surface decreases. By considering the three regions of the wavelength range, a better understanding of light reflection from the surface of papers with different degrees of aging can be obtained. These three ranges are as follows: 400–600 nm, 600–1000 nm, and 1000–1100 nm. In the first and third areas, considerable noise was observed, which seems inappropriate for examining the conditions of the paper samples and does not provide accurate information.

The second region is a significant part to monitor the paper status, as we can understand the impacts of paper thickness, type of oil, and aging degree. In Fig. 3, in general, the reflectance of the fresh paper surface is higher than that of other aged papers, and the most aged paper (6w) has the lowest values. In terms of thickness, a greater thickness (0.25 mm) has a higher reflectance; for example, the reflectance is nearly 100% at a

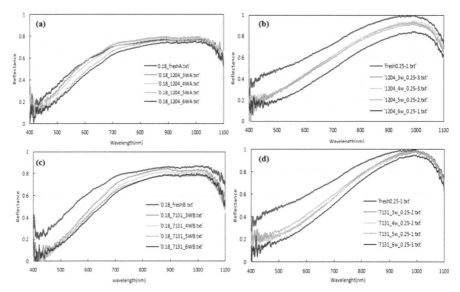

Fig. 3. Reflectance spectra response of aged paper samples. (a) Reflectance spectra of paper with a thickness of 0.18 mm in Natural Ester. (b) Reflectance spectra of paper with a thickness of 0.25 mm in Natural Ester. (c) Reflectance spectra of paper with a thickness of 0.18 mm in Synthetic Ester. (d) Reflectance spectra of paper with a thickness of 0.25 mm in Synthetic Ester.

wavelength of 1000 nm; however, it is less than 90% for a thinner thickness (0.18 mm). The effect of oil type on the reflectance is shown in Fig. 5 and will be discussed further. It would be better to mention that the 0.4 to 0.7 m and 0.7 to 2.5 m regions of the electromagnetic spectrum are visible and near-infrared regions.

Figure 4 shows the correlation between the DP values and the reflectance of the paper surfaces. For this purpose, three commercial wavelengths 635 nm, 808 nm, and 980 nm) were chosen. For Natural Ester oil (Fig. 4-a) a powerful correlation with $R^2 = 0.95$ was recorded for paper samples of the thickness of 0.18 mm at 980 nm. In addition, at the same wavelength, a correlation coefficient of $R^2 = 0.74$ was associated with the paper samples with a thickness of 0.25 mm. Therefore, strong correlation coefficients were found for thinner thicknesses at long wavelengths of 808 nm and 980 nm, and for greater thicknesses at short wavelengths of 635 nm. For Synthetic Ester oil (Fig. 4-b) a powerful correlation with $R^2 = 0.91$ was accorded with paper samples of the thickness of 0.25 mm at 635 nm. Furthermore, a poor correlation coefficient was related to the paper samples with a thickness of 0.18 mm at 980 nm. Therefore, in this type of oil, as shown in Fig. 4-b, as the wavelength increases, the strength of the correlation coefficient decreases, which applies to both thicknesses.

According to the correlation coefficient, it can be shown that there is a direct relationship between DP values and reflectance. A linear regression model analysis of a typical set of data made it possible to easily determine the DP value through this correlation. Figure 4 shows the correlation between the reflectance spectra and DP values for aged papers of two different thicknesses in two different oils from the ester fluid series.

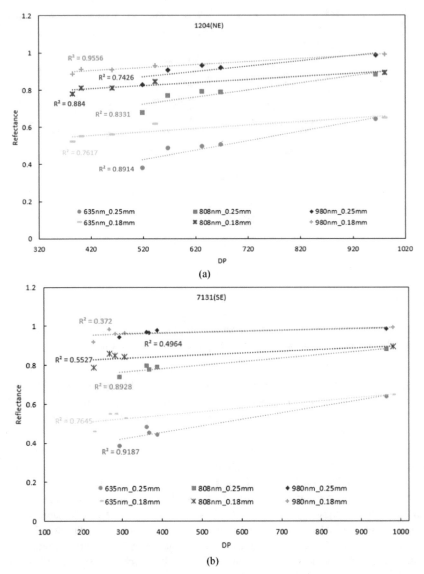

Fig. 4. Fitting line as a linear explaining the relation between DP and reflectance for three different wavelengths 635, 808, and 980 nm. (a) Correlation of reflectance and DP for papers in Natural Ester (1204). (b) Correlation of reflectance and DP for papers in Synthetic Ester (Midel 7131).

Figure 5 shows the effect of the type of oils and wavelengths on the reflection from the paper surface. The results indicate that the paper exhibits a reflectance of over 80% at longer wavelengths (980 nm), whereas it drops to below 60% at shorter wavelengths (635 nm). It is also observed that with aging, the decline in the reflectance is more consistent for paper aged in natural esters. Additionally, the DP value demonstrates that paper impregnated in natural ester oil has a greater degree of mechanical stability compared to paper aged in synthetic ester oil, as illustrated in Fig. 1.

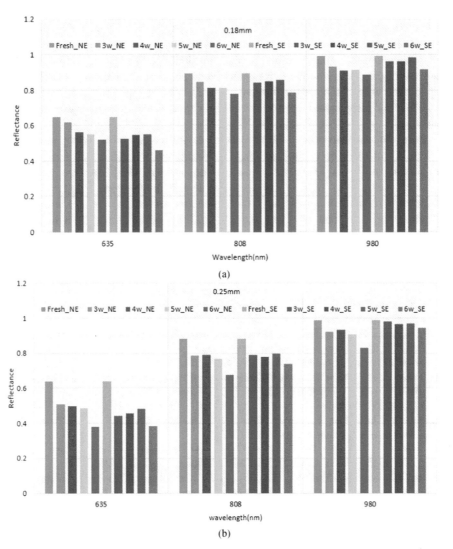

Fig. 5. The reflectance function of specific wavelengths in Ester fluids. (a) thickness 0.18 mm and (b) thickness of 0.25 mm.

4 Conclusion

This study found that there is a correlation between the degree of polymerization (DP) value and an optical technique. The relationship between the physical and mechanical characteristics of paper samples and their optical features is demonstrated by the correlation between DP and reflectance. The optical approach provides a quick and reliable measurement of the aging condition of paper insulation, offering a significant improvement over the traditional offline viscometric approach described in ASTM D4243. The experiments were conducted for different types of insulating liquid used, and these results provide initial insights into the potential of using optical techniques to determine the DP value.

Going forward, an effort will be made to create a centralized remote monitoring system that utilizes state-of-the-art telecommunications technologies such as sensors and key Internet of Things (IoT) communication protocols. The MQTT protocol, which is a message transfer protocol, will play a crucial role in remotely monitoring the condition of the paper insulation by sending data to a central server. This system will provide real-time monitoring and improve the overall efficiency of monitoring the aging process of paper insulation in oil-filled transformers.

References

1. Fofana, I., Hadjadj, Y.: Electrical-based diagnostic techniques for assessing insulation condition in aged transformers. Energies **9**(9), 679 (2016)
2. Loiselle, L., Fofana, I., Sabau, J., Magdaleno-Adame, S., Olivares-Galvan, J.C.: Comparative studies of the stability of various fluids under electrical discharge and thermal stresses. IEEE Trans. Dielectr. Electr. Insul. **22**(5), 2491–2499 (2015)
3. Ansari, M.A., Martin, D., Saha, T.K.: Investigation of distributed moisture and temperature measurements in transformers using fiber optics sensors. IEEE Trans. Power Delivery **34**(4), 1776–1784 (2019)
4. Simplice, A., Behjat, V., Kung, P., Fofana, I.: Assessing Water Content and Vibration from Dynamic Measurement in Transformer (2019)
5. Razzaq, A., Zainuddin, H., Hanaffi, F., Chyad, R.M.: Transformer oil diagnostic by using an optical fibre system: a review. IET Sci. Meas. Technol. **13**(5), 615–621 (2019)
6. Wang, L., Tang, C., Zhou, S., Wang, Q., Liu, X.: A novel method for the deterioration state evaluation of mineral insulating oil by THz time-domain spectroscopy. IEEE Access **7**, 71167–71173 (2019)
7. Hayber, ŞE., Tabaru, T.E., Güçyetmez, M.: Evanescent field absorption-based fiber optic sensor for detecting power transformer oil degradation. Fiber Integr. Opt. **40**(4–6), 229–248 (2021)
8. Li, S., Ge, Z., Abu-Siada, A., Yang, L., Li, S., Wakimoto, K.: A new technique to estimate the degree of polymerization of insulation paper using multiple aging parameters of transformer oil. IEEE Access **7**, 157471–157479 (2019)
9. Arroyo, O.H., Fofana, I., Jalbert, J., Ryadi, M.: Relationships between methanol marker and mechanical performance of electrical insulation papers for power transformers under accelerated thermal aging. IEEE Trans. Dielectr. Electr. Insul. **22**(6), 3625–3632 (2015)
10. Rozga, P., Stanek, M., Pasternak, B.: Characteristics of negative streamer development in ester liquids and mineral oil in a point-to-sphere electrode system with a pressboard barrier. Energies **11**(5), 1088 (2018)

11. N'cho, J.S., Fofana, I.: Review of fiber optic diagnostic techniques for power transformers. Energies **13**(7), 1789 (2020)
12. Gupta, B.: Direct determination of moisture in solid oil-paper insulation, pp. 583–586
13. Medina, R.D., Romero, A.A., Mombello, E.E., Ratta, G.: Assessing degradation of power transformer solid insulation considering thermal stress and moisture variation. Electr. Power Syst. Res. **151**, 1–11 (2017)
14. Baird, P.J., Herman, H., Stevens, G.C., Jarman, P.N.: Non-destructive measurement of the degradation of transformer insulating paper. IEEE Trans. Dielectr. Electr. Insul. **13**(2), 309–318 (2006)
15. Norazhar, A.B., Abu-Siada, A., Islam, S.: A review on chemical diagnosis techniques for transformer paper insulation degradation. pp. 1–6
16. Bakar, N.A., Abu-Siada, A., Islam, S.: A review of dissolved gas analysis measurement and interpretation techniques. IEEE Electr. Insul. Mag. **30**(3), 39–49 (2014)
17. Thiviyanathan, V.A., Ker, P.J., Leong, Y.S., Abdullah, F., Ismail, A., Jamaludin, M.Z.: Power transformer insulation system: a review on the reactions, fault detection, challenges and future prospects. Alexandria Eng. J. **61**(10), 7696–7713 (2022)
18. Mahanta, D.K., Laskar, S.: Transformer condition monitoring using fiber optic sensors: a review. ADBU J. Eng. Technol. **4** (2016). ISSN 2348-7305
19. Zhang, W.-C., Chen, Q.-C., Zhao, H.: Numerical investigation of acoustic emissions distribution from partial discharge in transformer, pp. 498–501 (2017)
20. Münster, T., Werle, P., Peter, I., Hämel, K., Barden, R., Preusel, J.: Optical sensor for determining the degree of polymerization of the insulation paper inside transformers. Transformers Mag. **8**(3), 106–117 (2021)
21. Zeng, W., Gao, H.: Optic fiber sensing IOT technology and application research. Sens. Transducers **180**(10), 16 (2014)
22. Zhang, R., Yan, Z., Tong, J.: Fiber grating sensor based Internet of Things for intelligent substations, pp. 419–422
23. Chehri, A., Fofana, I., Yang, X.: Security risk modeling in smart grid critical infrastructures in the era of big data and artificial intelligence. Sustainability **13**(6), 3196 (2021)
24. Wang, W., He, D., Gu, J., Lu, J., Du, J.: Electrical–thermal aging characteristic research of polymer materials by infrared spectroscopy. Polym. Adv. Technol. **25**(12), 1396–1405 (2014)

Digital Enterprise Architecture
for Human-Centric Intelligent Systems
in Manufacturing, Financial, and Others

ChatGPT, How to Wire Age 5.0 Mindsets: Industry, Society, Healthcare and Education?

Abdellah Chehri[1]($\boxtimes$), Hasna Chaibi[2], Alfred Zimmermann[3], and Rachid Saadane[4]

[1] Department of Mathematics and Computer Science, Royal Military College of Canada, Kingston, ON 11 K7K 7B4, Canada
chehri@rmc.ca

[2] GENIUS Laboratory, SUPMTI Rabat, Casablanca, Morocco

[3] Reutlingen University, Reutlingen, Germany
alfred.zimmermann@reutlingen-university.de

[4] SIRC-(LaGeS), Hassania School of Public Works, Casablanca, Morocco
saadane@ehtp.ac.ma

Abstract. In today's education, healthcare, and manufacturing sectors, organizations and information societies are discussing new enhancements to corporate structure and process efficiency using digital platforms. These enhancements can be achieved using digital tools. Industry 5.0 and Society 5.0 give several potentials for businesses to enhance the adaptability and efficacy of their industrial processes, paving the door for developing new business models facilitated by digital platforms. Society 5.0 can contribute to a super-intelligent society that includes the healthcare industry. In the past decade, the Internet of Things, Big Data Analytics, Neural Networks, Deep Learning, and Artificial Intelligence (AI) have revolutionized our approach to various job sectors, from manufacturing and finance to consumer products. AI is developing quickly and efficiently. We have heard of the latest artificial intelligence chatbot, ChatGPT. OpenAI created this, which has taken the internet by storm. We tested the effectiveness of a considerable language model referred to as ChatGPT on four critical questions concerning "Society 5.0", "Healthcare 5.0", "Industry," and "Future Education" from the perspectives of Age 5.0.

Keywords: ChatGPT · Smart City · Society 5.0 · Industry 5.0 · Healthcare 5.0

1 Introduction

In the most recent decades of human history, technological development and growth have made extraordinary leaps and bounds, particularly with the advent of the Internet [1–4]. The world is currently in a new era, in which globalization and the rapid evolution of digital technologies such as the Internet of Things (IoT), Big Data (BD), Artificial Intelligence (AI), robotics, 3D printing, Cloud Computing (CC), and Mobile Devices (MD), amongst others, are pushing for major changes in business and society, and creating an entirely new environment. This new era has brought about a new environment for the world.

© The Author(s), under exclusive license to Springer Nature Singapore Pte Ltd. 2023
A. Zimmermann et al. (Eds.): KES-HCIS 2023, SIST 359, pp. 133–142, 2023.
https://doi.org/10.1007/978-981-99-3424-9_13

The term "Industry 4.0" was introduced by a German research institute in 2011 [5]. The German government initiated a discussion on the manufacturing sector's future, headed by the country's intellectual community and key industrial partners. For want of a better word, the goal was to pinpoint the exact conditions under which Germany's manufacturing sector would become the world's most productive and adaptable [6].

The 4.0 Industrial Revolution is affecting all sectors of the economy, including the agricultural sector [7–10], smart water management [11] and is reshaping their production capacity.

It is Japan's ambition to be the first nation in the world to realize a human-centered society (also known as Society 5.0), which is a society in which everyone has the opportunity to live a life of high quality and full of vitality. It plans to achieve this goal by integrating cutting-edge technologies into a wide variety of economic sectors and social endeavors and stimulating innovation to generate new forms of value [12]. Therefore, the advancement of human civilization is connected to the continuously shifting economic formations, and the current social and economic condition is determined by notions such as Society 5.0 and the fourth and fifth industrial revolutions.

The advancements that have been made in neural networks, deep learning, and artificial intelligence (AI) over the last decade have revolutionized how we approach a wide variety of jobs and industries, including manufacturing, banking, and consumer products. This has caused a sea change in how we think about these fields [13]. Understanding the new realities brought about by globalization, cultural shifts, and the proliferation of information and communication technologies (ICT) calls for four mindsets (industry, society, health and education) to converge in order to "learn, think, research, create, and change," as shown in Fig. 1.

Fig. 1. The five major tools within the four mindsets of Age 5.0.

The ultimate "change" defines those future institutions of higher learning that must adapt along with the populations they serve to remain competitive and ready to account for their decisions to a wide range of stakeholders.

The capacity to construct highly accurate classification models rapidly and independently of the type of input data (for example, images, text, or audio) has made it possible for widespread adoption of applications such as automated tagging of objects and users in photographs, near-human level text translation, automatic scanning in bank ATMs, and even the generation of image captions. These applications have been made possible by the ability to build highly accurate classification models.

Many people are testing out ChatGPT and writing about their experiences with it on social media, so it's become a topic that's getting a lot of attention in the news and online. This cutting-edge technology, which OpenAI created, has attracted the attention of people worldwide due to its extraordinary intelligence.

ChatGPT, which was developed by OpenAI and is based in San Francisco, California, is a significant language model that generates natural language responses to the text input in the context of a conversation by using self-attention mechanisms and a considerable quantity of training data. It is beneficial for managing dependencies over a lengthy period of time and producing responses that are consistent with the context in which they are being applied.

The ChatGPT language model is entirely enclosed within the server. Hence it is unable to surf the internet or does web searches. As a result, the neural network generates all responses "in situ," depending on the conceptual relationship that exists between individual words (also known as "tokens"). This is in contrast to other chatbots or conversational systems, which are granted permission to access external sources of information to deliver directed responses to user inquiries (for example, by running web searches or accessing databases). The following is the order in which the questions were entered into ChatGPT after being formatted into their respective variants:

What exactly is the 4.0 Industry?
What obstacles must be overcome in order to implement industry 4.0?
What exactly is this "Society 5.0"?
What will the future of healthcare be like in the years to come?

To the best of our knowledge, this is the first paper that evaluates the accuracy and concordance of ChatGPT for four mindsets (industry, society, health, and education) from the perspectives of Age 5.0.

The following describes the format of this paper: Section 2 is related to Industry 4.0. The description of Society 5.0 is given in Sect. 3. Section 3 explains how advanced data analytics systems for healthcare 5.0. In Sect. 4, we discuss the future of education in Age 5.0. The conclusion and the article's perspective could be addressed in Sect. 5.

2   ChatGPT What is Industry 4.0?

Because Industry 4.0 does not have a singular definition, the technologies connected with it cannot be identified with absolute certainty [14–15]. Autonomous robots, simulation, horizontal and vertical integration of systems, the Internet of Things, the cloud, additive manufacturing, augmented reality, Big Data, and cybersecurity are the nine primary technologies that have been identified as being associated with Industry 4.0. Every one of the technological advancements envisioned by the concept of Industry 4.0 enhances several facets of production, including the planning of operations, the maintenance of equipment, and the management of inventory (Fig. 2).

However, for businesses to raise their profits and levels of production, they have frequently concentrated their efforts on implementing technology associated with Industry 4.0. In doing so, they have occasionally neglected the human element. If the implementation of these new technologies impacts it, then it should be no surprise that it also has an effect on these techniques in turn. When the performance of these technologies leads to an improvement in the execution of operations while simultaneously offering expected and appropriate support to men in carrying out the activities for which they are responsible, this feedback is shown to be beneficial. This holds when it comes to making decisions in the context of operational settings.

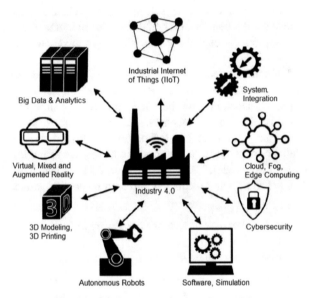

Fig. 2. Major technologies in Industry 4.0.

But how the ChatGPT thinks? If we ask the question "what is Industry 4.0", the results are pretty accurate. Below is the chatGPT answer.

CHatGPT's Answer: *"Industry 4.0 is the current state of the industrial sector, where automation and data exchange are used to improve the efficiency of factories. Industry 4.0 is the next step, where cyber-physical systems are used to create a "smart factory" with a high degree of automation. This includes the use of robots, 3D printing, and the internet of things."*

On the other hand, by doing some literature review, we could define the challenge related to Industry 4.0. According to ChatGPT, six obstacles are defined by ChatGPT, when asking the following question to ChatGPT, the answer is more or less accurate.

What obstacles must be overcome in order to implement industry 4.0?

1. Cost: Implementing Industry 4.0 technologies and processes can be expensive. Companies need to ensure that the benefits outweigh the costs.

2. Security: Industry 4.0 brings with it a new set of cyber security risks. Companies need to ensure that their systems are secure and that data is protected.
3. Education: Companies need to ensure that their staff are educated and trained in the use of Industry 4.0 technologies.
4. Legislation: Companies need to be aware of the relevant legislation and regulations that apply to Industry 4.0.
5. Talent: Companies need to ensure that they have the right talent to make best use of Industry 4.0 technologies.
6. Scalability: Companies need to ensure that their systems are able to scale up to meet the demands of Industry 4.0.

3   Society 5.0

Data are gathered from the "real world" and fed into computers, where they are then processed, and the findings are then implemented in the "real world." This is the fundamental concept behind Society 5.0. In and of itself, this model is not novel. To use a paradigm that most people are familiar with, air conditioning machines automatically maintain a room at the temperature that has been set on the device. The temperature of the room is measured on a regular basis by an air conditioner, and an internal microcomputer then evaluates the reading in relation to the temperature that was previously registered.

In order to accomplish this goal, the Japan Business Federation (Keidanren) has released a vision paper (also known as an "outline") in which it explains that overcoming the problems that we have discussed, as well as other challenges, will require the destruction of five walls. Indeed, five, as in the 5.0 version of Society.

However, just as Industry 4.0 is depicted as the fourth industrial revolution in the position paper of Keidanren, so too is Society 5.0 described as an evolution in five societal stages (Table 1):

Table 1. Contextualizing Society 5.0. Modified from a Research conducted by the University of Tokyo's Material Innovation Research Center [12].

	Society 1.0	Society 2.0	Society 3.0	Society 4.0	Society 5.0
Society	Hunter-gatherer	Agrarian	Industrial	Information	Super smart
Industry	Capture/Gather	Manufacture	Mechanization	ICT	Merging of cyberspace and physical space
Material	Stone · Soil	Metal	Plastic	Semiconductor	Material 5.0*
Transport	Foot	horse	Motor car, boat, train plane	Multimobility	Autonomous driving
City ideals	Viability	Defensiveness	Functionality	Profitability	Humanity

In Society 5.0, AI analysis of big data in a database including varied sorts of information, such as sensor data from autos, real-time information on the weather, traffic, lodgings, and food and beverage, and personal history, will provide the following new forms of value [16–17].

4   ChatGPT What is Healthcare 5.0?

For many years, computer and communication technologies have accompanied the care process (diagnostic and therapy), giving rise to "trends" that have been dubbed "medicine 2.0" and "medicine 3.0," respectively. Artificial intelligence, learning algorithms (automated and deep), and the Internet of Things all point to the eventual arrival of "medicine 4.0," in which machines will play an increasingly significant role in diagnostic, therapeutic, and even preventive care. The future Medicine will be cutting-edge and high-performance, the outcome of tight collaboration between the neural brain and the silicon brain.

4.1   Healthcare 5.0

Healthcare in the future will be more preventive than curative. We will be able to recognize the disease early, intervene proactively, and understand its progression thanks to advances in science, big data, and technology [18–22]. Since the end of the nineteenth century, Medicine has evolved. It has evolved from merely curative therapy to holistic medicine. The Medicine of the twenty-first century will be so precise that it will drastically alter the care paradigm; it will be accurate, anticipatory, intelligent, networked, robotic, and digitalized. Above all, healthcare 5.0 will be called Healthcare P4 (Personalized, Predictive, Preventive, and Participatory). The aspects described by the term P4 Medicine summarize the functional framework of personalized medicine globally (Fig. 3).

- **Personalized Medicine:** The concept of personalized Medicine is not new; physicians have always strived to tailor care to better adapt it to people's particular health requirements (each doctor has always aimed to give the right Medicine at the right time and to the right patient). Throughout the history of Medicine, it has never been able to correctly forecast each body's reaction to specific interventions or identify the amount of risk at which we can develop a disease.
- **Predictive Medicine:** The term "predictive" or "forecasting" medicine refers to therapy based on expectation and targeted not at patients but at healthy individuals at risk of developing a particular condition. To begin with, predictive Medicine differs from preventive Medicine, even though their goals are identical. In Predictive Medicine, the description of pharmaceuticals is mostly based on patient-specific data, particularly critical characteristics, which may include genetic information. This conclusion is beneficial for evaluating the adequacy of proposed therapies and avoiding all outcomes and undesirable reactions.
- **Preventive Medicine:** Unlike predictive Medicine, which is primarily probabilistic, preventive Medicine is more concrete and traditional, relying on health education, awareness-raising, and even IEC (Information-Education-Communication)

programs aimed at reducing disease risks (primary prevention), treating "clinical" disease as early as possible through early detection (secondary prevention) and constantly seeking to improve the quality of life of people who are already sick (tertiary prevention).

- **Participatory Medicine:** This facet involves patient participation in managing personal data. In addition to their experiences and self-monitoring of their lives and relevant events, this treatment is based on their theoretical medical knowledge. In recent years, Medicine has become increasingly participatory, with healthcare providers and patients working in collaboration and "participation". The care process increasingly incorporates "a patient dimension," a digital representation or digital base combining the personalized information extracted or built from his body and lifestyle. For his part, the doctor is supported by his knowledge, experience, and decision-making rules. With the technological development of health systems that have become more efficient in terms of remote patient control and monitoring, this development facilitates the participation of patients through their information based on connected electronic sensors. It enables more accurate tracking of individuals' daily habits, such as sleep, nutrition, and physical activity. It allows better monitoring of the daily routines of individuals, such as sleep, nutrition, sports exercises, and other lifestyles (genetic heritage, family history, travel, pollution level…).

Fig. 3. The four aspects of 4P medicine.

5   Education 5.0

Since the 1990s, with the advent of information and communication technologies (ICT), we have witnessed the virtualization of educational technology (commonly abbreviated as EduTech, or EdTech). Today, with the aid of technology, we can organise education in ways that were previously unimaginable. It has various advantages, including cost-effectiveness, greater reach, scalability, and adaptability. When COVID-19 hit the globe, AI, robotics, Big Data, and other technological advancements were already fundamentally and rapidly altering modern education [23, 24].

To adapt education to the requirements of the Fourth Industrial Revolution, however, may already be a case of falling behind the train. Experts are increasingly proposing the concept of the Fifth Industrial Revolution (Industry 5.0).

A comprehensive educational shift is necessary in order to reach Education 5.0, and this shift necessitates that all pertinent aspects of education be addressed [25]. In addition to the technological aspects, the following aspects also need to be addressed:

- **Strategy:** the strategy will consist of reframing the primary purpose and particular goals of educational opportunities within the context of Education 5.0;
- **Collaboration:** fostering behaviors that go beyond the normal institutional collaboration patterns and involving individuals and communities, as well as specifically developing effective learning ecosystems that engage all important stakeholder groups;
- **Material:** identifying, producing, and introducing content that corresponds to the Strategy element (including maintaining a good balance of technical and non-technical disciplines, paying special attention to the concerns of ethics, social inclusion, diversity, and sustainability, etc.);
- **Learning environment:** the creation of a learning environment that best serves the specific objectives of the Strategy element (for example, methods that stimulate multidisciplinary orientation, design thinking, team spirit, collective problem-solving, risk-taking behaviour, experimental approaches, and so on);
- **Delivery mechanisms:** determining which tools are best suitable for fulfilling the objectives of the Strategy element; this is the point at which technology may or may not be selected as the most effective delivery mechanism.
- **Quality Assurance:** developing defined quality requirements for Education 5.0 and carrying out ongoing quality monitoring.

6   Conclusion

While other generative AI systems have recently been accessible, ChatGPT is now the most well-known and has received extensive media coverage. This artificial intelligence application uses predictive technology to generate or update all forms of textual output, including computer code, work plans, articles (such to this one), and reports.

As ChatGPT users and clients, our daily routines, working environments, and team relationships may be altered. Could it accelerate some of the monotonous tasks we abhor? Could it bring greater depth and originality to our work plans? Could it aid in skill development and education? These and many questions inspired me to write this article. The tool normally produces high-quality outputs, despite the possibility of errors.

References

1. Masuda, Y., Zimmermann, A., Shepard, D.S., Schmidt, R., Shirasaka, S.: An Adaptive Enterprise Architecture Design for a Digital Healthcare Platform: Toward Digitized Society – Industry 4.0, Society 5.0. In: IEEE 25th International Enterprise Distributed Object Computing Workshop (EDOCW). Gold Coast, Australia 2021, pp. 138–146 (2021)

2. Schmidt, R., Zimmermann, A., Keller, B., Möhring, M., "Towards Engineering Artificial Intelligence-based Applications,",: IEEE 24th International Enterprise Distributed Object Computing Workshop (EDOCW). Eindhoven, Netherlands **2020**, 54–62 (2020)
3. Fritzsch, J., Bogner, J., Wagner, S., Zimmermann, A.: Microservices Migration in Industry: Intentions, Strategies, and Challenges. In: 2019 IEEE International Conference on Software Maintenance and Evolution (ICSME), Cleveland, OH, USA, pp. 481–490 (2019)
4. Bogner, J., Fritzsch, J., Wagner, S., Zimmermann, A.: Microservices in Industry: Insights into Technologies, Characteristics, and Software Quality. In: 2019 IEEE International Conference on Software Architecture Companion, Hamburg, Germany, pp. 187–195 (2019)
5. Chehri, A., Zimmermann, A., Schmidt, R., Masuda, Y.: Theory and practice of implementing a successful enterprise IoT strategy in the industry 4.0 era. Procedia Comput. Sci. **192**, 4609–4618 (2021)
6. Zimmermann, A., Schmidt, R., Sandkuhl, K., Masuda, Y., Chehri, A.: Architecting Intelligent Service Ecosystems: Perspectives, Frameworks, and Practices. In: Buchmann, R.A., Polini, A., Johansson, B., Karagiannis, D. (eds.) BIR 2021. LNBIP, vol. 430, pp. 150–164. Springer, Cham (2021). https://doi.org/10.1007/978-3-030-87205-2_10
7. Ouafiq, E.M., Elrharras, A., Mehdary, A., Chehri, A., Saadane, R., Wahbi, M.: IoT in Smart Farming Analytics, Big Data Based Architecture. In: Zimmermann, A., Howlett, R.J., Jain, L.C. (eds.) Human Centred Intelligent Systems. SIST, vol. 189, pp. 269–279. Springer, Singapore (2021). https://doi.org/10.1007/978-981-15-5784-2_22
8. Chehri, A., Chaibi, H., Saadane, R., Hakem, N., Wahbi, M.: A framework of optimizing the deployment of IoT for precision agriculture industry. Procedia Comput. Sci. **176**, 2414–2422 (2020)
9. Ouafiq, E.M., Saadane, R., Chehri, A.: Data management and integration of low power consumption embedded devices IoT for transforming smart agriculture into actionable knowledge. Agriculture **12**, 329 (2022). https://doi.org/10.3390/agriculture12030329
10. Ouafiq, E.M., Saadane, R., Chehri, A., Jeon, S.: AI-based modeling and data-driven evaluation for smart farming-oriented big data architecture using IoT with energy harvesting capabilities. Sustain. Energy Technol. Assess. **52**, 102093 (2022)
11. Quadar, N., Chehri, A., Jeon, G., Ahmad, A.: Smart Water Distribution System Based on IoT Networks, a Critical Review. In: Zimmermann, A., Howlett, R.J., Jain, L.C. (eds.) Human Centred Intelligent Systems. SIST, vol. 189, pp. 293–303. Springer, Singapore (2021). https://doi.org/10.1007/978-981-15-5784-2_24
12. Hitachi-U Tokyo Laboratory. Society 5.0: a people-centric super-smart society. Singapore: Springer (2020)
13. Habash, R.: Phenomenon-based Learning for Age 5.0 Mindsets: Industry, society, and Education. In: IEEE Global Engineering Education Conference, Tunisia, pp. 1910–1915 (2022)
14. Chehri, A., Jeon, G.: The Industrial Internet of Things: Examining How the IIoT Will Improve the Predictive Maintenance. In: Chen, Y.-W., Zimmermann, A., Howlett, R.J., Jain, L.C. (eds.) Innovation in Medicine and Healthcare Systems, and Multimedia. SIST, vol. 145, pp. 517–527. Springer, Singapore (2019). https://doi.org/10.1007/978-981-13-8566-7_47
15. Chehri, A., Fortier, P., Tardif, P.: An investigation of UWB-based wireless networks in industrial automation. Int. J. Comput. Sci. Netw. Security **8**, 179–188 (2008)
16. Sharma, T., Debaque, B., Duclos, N., Chehri, A., Kinder, B., Fortier, P.: Deep learning-based object detection and scene perception under bad weather conditions. Electronics **11**, 563 (2022). https://doi.org/10.3390/electronics11040563
17. Ahmed, I., Jeon, G., Chehri, A., Hassan, M.M.: Adapting Gaussian YOLOv3 with transfer learning for overhead view human detection in smart cities and societies. Sustain. Cities Soc. **70**, 102908 (2021)

18. Chehri, A., Mouftah, H.T.: Internet of things - integrated IR-UWB technology for healthcare applications. Concurr. Computat Pract Exper. **32**, e5454 (2020). https://doi.org/10.1002/cpe.5454
19. Chehri, A., Moutah, H.T.: Survivable and Scalable Wireless Solution for E-health and E-emergency Applications. In Proceedings of the EICS4Med, Pisa, Italy, 13–16, pp. 25–29 (2011)
20. Chehri, A.: Energy-efficient modified DCC-MAC protocol for IoT in e-health applications. Internet Things **14**, 100119 (2021)
21. Ahmed, I., Jeon, G., Chehri, A.: An IoT-enabled smart health care system for screening of COVID-19 with multi layers features fusion and selection. Computing , 1–18 (2021). https://doi.org/10.1007/s00607-021-00992-0
22. Ahmed, I., Chehri, A., Jeon, G.: A Sustainable deep learning-based framework for automated segmentation of COVID-19 infected regions: using U-Net with an attention mechanism and boundary loss function. Electronics **11**, 2296 (2022)
23. Chehri, A., Popova, T.N., Vinogradova, N.V., Burenina, V.I.: Use of Innovation and Emerging Technologies to Address Covid-19-Like Pandemics Challenges in Education Systems. In: Uskov, V.L., Howlett, R.J., Jain, L.C. (eds.) Smart Education and e-Learning 2021. SIST, vol. 240, pp. 441–450. Springer, Singapore (2021). https://doi.org/10.1007/978-981-16-2834-4_38
24. Mitrofanova, Y.S., Chehri, A., Tukshumskaya, A.V., Vereshchak, S.B., Popova, T.N.: Project Management of Smart University Development: Models and Tools. In: Uskov, V.L., Howlett, R.J., Jain, L.C. (eds.) Smart Education and e-Learning 2021. SIST, vol. 240, pp. 339–350. Springer, Singapore (2021). https://doi.org/10.1007/978-981-16-2834-4_29
25. Dervojeda, K.: Education 5.0: focus on pupils, not on technology, PwC, 2022, https://www.pwc.nl/en/insights-and-publications/services-and-industries/public-sector/education-focus-on-pupils-not-on-technology.html

Designing Performance Indicator in Human-Centered Agile Development

Kasei Miura[1(✉)], Yoshimasa Masuda[1,2], and Seiko Shirasaka[1]

[1] Graduate School of System Design and Management, Keio University, Kanagawa, Japan
kasei.miura@keio.jp

[2] The School of Computer Science, Carnegie Mellon University, Pittsburgh, PA, USA

Abstract. Advances in information technology are transforming the value provided to customers and business models for delivering value. Pharmaceutical companies, which are required to provide safe and reliable products, are also building digital platforms to comply with regulatory and user requirements and to improve the efficiency of manufacturing and quality control. In order to be accountable for the reliability of their products and services, they need to continuously assess their performance in the process of digital transformation.

However, there is no established method to dynamically design and manage performance indicators in the process from the verification of concept of a new product to its deployment to the business. In this report, we propose a method for deriving dynamic performance indicators for corporate digital transformation using a design thinking approach. This method enables more dynamic and flexible design and review of performance indicators, instead of the conventional static performance indicators centered on financial perspectives.

Keywords: Performance Indicator · Adaptive Integrated Digital Architecture Framework · Agile Development · Design Thinking

1 Introduction

Today's pharmaceutical companies are considering new approaches to manufacturing, quality, and supply chain efficiency through digital platforms such as IoT, digital twin, blockchain, big data, and AI. IoT can be used to monitor facilities and equipment and factory staffs behavior on the production floor and to simulate production management with a digital twin [1, 2]. In addition, blockchain technology is expected to assist in understanding the traceability of supply chains. Furthermore, big data and AI are expected to improve the accuracy of demand forecasting [3]. Industry 4.0 offers opportunities for companies to work with the ecosystem to increase the flexibility and efficiency of their supply chains [4]. In the pharmaceutical industry, there is an increasing need to use new production technologies, respond to changing customer needs, and work more closely with the ecosystem. As a result, digital transformation is being considered to increase the efficiency of production processes without compromising product reliability. In addition, digital transformation is expected to contribute to sustainability and resilience in the supply chain by enhancing the capability to monitor, respond, learn, and anticipate.

© The Author(s), under exclusive license to Springer Nature Singapore Pte Ltd. 2023
A. Zimmermann et al. (Eds.): KES-HCIS 2023, SIST 359, pp. 143–152, 2023.
https://doi.org/10.1007/978-981-99-3424-9_14

Performance evaluation is useful in the digital transformation of an organization to help clarify objectives, communicate them to stakeholders, and improve initiatives. Performance evaluation clarifies the components related to the organization's objectives and provide information for accountability and improvement regarding projects and initiatives. In addition, in an environment known as VUCA, performance evaluation must respond to changes in the environment, changes in objectives, and changes in the organization's strategy. In light of this situation, Miura et al. propose the Strategic Performance Indicator Derivation Framework (SPIDF) [5] (Fig. 1).

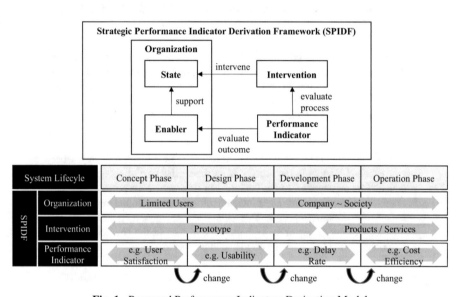

Fig. 1. Proposed Performance Indicators Derivation Model

User-centered agile development methodologies can reduce risk by clarifying the value to be delivered in digital transformation, starting with proof-of-concept and prototyping, and then scaling up. However, existing performance evaluation approaches are not sufficient to address design thinking and agile development approaches in the above digital transformation process. In the case of waterfall development, performance indicators can be handled through static management, whereas in the agile approach, the functions and values provided by the system are continuously changing, and dynamic performance evaluation is required to respond to these changes.

Enterprise architecture frameworks are helping to drive the digital transformation of the enterprise [6]. It is a systematic approach that provides EA helps create a roadmap from current architecture to future architecture [7]. In addition, an adaptive integrated digital framework has been proposed and is expected to be used to drive digital transformation [8]. Masuda et al. investigate the architectural design and implementation of digital healthcare platforms in the healthcare community and propose AIDAF. In addition, they review case studies where digital platforms were designed and built using design thinking and agile software development methods. The AIDAF is then proposed and validated in conjunction with the Design Thinking approach [9]. A quality management system is a comprehensive system that oversees all activities and deliverables of a company related to quality. Since departments with their own organizational strategies need to be managed in an integrated manner, the application of the Adaptive Integrated Digital Architecture Framework (AIDAF) is expected to be effective.

This study proposes the use of SPIDF in conjunction with AIDAF when applying an agile development approach. Specifically, the research questions employed in this study are as follows.

RQ1: How can the proposed performance indicator derivation method in pharmaceutical supply chain, manufacturing, and quality support to design performance indicators in design thinking and agile development approaches?

RQ2: How can performance indicators be managed and improved during the digital transformation process?

This paper is organized as follows. Provides the background of this study and outlines the proposed application and application of performance indicators. Discusses the application of the performance indicator derivation method in the design thinking approach. Propose the purpose and method of setting performance indicators according to the situation in each phase of development. Then, future directions for the validation of this research will be discussed.

2   Related Research

2.1   Performance Evaluation

Performance evaluation plays an important role in materializing strategies, communicating them to stakeholders, and improving their effectiveness [10]. As a method for organizations to comprehensively evaluate business performance, a method that systematically uses performance indicators in combination with leading indicators, in addition to conventional financial indicators, has been proposed and used. Balanced scorecard provides four perspectives for overarching measurement: financial, customer, process, and learning and growth [11]. An organization's strategy is dynamically managed in response to changes in the organization's environment [12]. However, a divergence has emerged between the organization's strategy and the management of performance indicators [13]. Another approach to evaluating performance with a focus on strategic intervention is program evaluation [14]. Program evaluation is used to assess goals and results to demonstrate accountability and to evaluate the implementation of measures to improve business. Program evaluation involves an assessment of needs and an assessment of program theory with respect to program evaluation. The results of that assessment are used to examine the process, outputs, and outcomes impact to be evaluated. The logic model is the methodology used in the program theory assessment. The logic model clarifies program inputs, activities, outputs, and outcome as hypotheses, and identifies critical evaluation points. In recent years, in response to the increasing volatility, uncertainty, complexity, and ambiguity of the organizational environment, organizations have been changing their strategies more dynamically to achieve their objectives. However, it is sometimes difficult to align performance indicators with changes in strategy, result-ing in inconsistencies, and methods are needed to manage performance measurement appropriately and dynamically.

Considering the above background, previous studies have proposed a framework for deriving performance indicators named as Strategic Performance Indicator Derivation Framework (SPIDF) based on the relationship between the organizational state and the interventions to achieve it [5]. This framework for deriving performance indicators consists of the following four steps; (1) Estimate the transition state of the organization, (2) Analyze the enablers of each state of organization, (3) Functional analysis of strategic interventions, (4) Derive performance indicators from the analysis of organizational enablers and strategic interventions.

It is suggested that this framework will allow timely measurement and evaluation of organizational performance. It is also suggested that a combination of AIDAF and SPIDF can be used to dynamically manage system planning through an adaptive EA lifecycle, utilizing existing EA that is appropriate for each department.

2.2 Quality Management System

In the healthcare industry, companies are accountable for the products they provide and are required to establish quality management systems to comply with regulatory and user requirements [15]. Quality management systems require various performance assessments to demonstrate their effectiveness, adequacy, and appropriateness [16]. The performance related to adequacy is required to evaluate the conformity to requirements, i.e., regulatory requirements. Performance related to effectiveness requires an assessment of whether or not the goals set forth in the quality policy have been achieved. In addi-tion, appropriateness is evaluated in terms of resource adequacy and cost-effectiveness. These performance indicators are monitored through the collection of external feedback information and process information, and are required to be analyzed and reviewed to ensure that the standards are met during the process of change in the quality manage-ment system [17, 18]. Total Quality Management is reported as a way to make quality management systems responsive to changes in the business environment and to achieve sustainable growth [19].

2.3 Adaptive Integrated Digital Architecture Framework

Digital technologies are supporting transformation in companies. Enterprises are using digital technologies to offer new value, improve operational efficiency, and create new business models. Enterprise architecture frameworks are being used to leverage digital technologies in transforming enterprise systems. The enterprise architecture framework must be effective in order to leverage digital technologies to solve organization-wide issues. Therefore, from a comprehensive perspective, it is structured to encompass all enterprise artifacts, including business, applications, data, and infrastructure, and to build a roadmap from current architecture to future architecture. On the other hand, the EA framework itself has also been studied to keep pace with the evolution of digital technologies. Prior research has proposed the Adaptive Integrated Digital Architecture Framework (AIDAF), which is aligned with the digital strategy [8]. In the adaptive integrated digital cycle that AIDAF presents, digitization projects are driven as short-term cycles. It begins with a context phase in which a digital technology implementation plan is developed according to the digital strategy. In the evaluation and architecture study phase, the alignment of the IT systems described in the implementation plan with the enterprise architecture is examined. In the rationalization phase, the applicability of the proposed new IT system is determined. In the realization phase, issues and initiatives in the IT system are deliberated, and the new IT project is initiated.

2.4 Design Thinking Approach

Companies sometimes fail in product development because they cannot provide the right product or service that users want. This is due to the gap between the company's assumed user needs and the true needs of users. Another factor is that users do not understand what they need. Therefore, the design thinking approach is utilized to capture the true needs of users [20]. The design thinking approach was proposed by Stanford University as one that should have four philosophies: (1) human-centered, (2) creative, (3) hands-on, and (4) iterative process [21]. The design thinking approach is a user-centered and creativity-driven approach to understanding the true needs of users. Understanding what users want is important, and ethnography and prototyping are used for this purpose. In prototyping, a product with minimal functionality is used to evaluate user satisfaction and demonstrate the product concept. In the design thinking approach, it is important to try iterative experiments to materialize the value provided.

2.5 Agile Development Approach

The development of digital technology has made people and things interconnected, and closed systems have become more open. This openness has led to greater uncertainty, as they are more susceptible to the influence of the external environment and their relationships with stakeholders. The increase in uncertainty is also accelerated by the rapid pace of technological development. Systems development has traditionally been conducted in a waterfall model [22]. In the waterfall model, requirements are defined at the start of a project based on a business case, design is developed, and then the project is implemented. In the waterfall model, the challenge is to reduce uncertainty. To reduce uncertainty, the size of deliverables to be created at one time is made smaller to reduce the risk in terms of customer value, quality, cost, and delivery time. For this reason, agile development is becoming increasingly popular. Agile-type development enables the delivery of necessary value while minimizing risk by reducing the size of the deliverable, and by making the evaluation and realization of customer value faster and more iterative [23]. Several agile development frameworks have been developed to date [24–26].

3 Proposal of Strategic Performance Indicator Deriving Framework for Design Thinking Approach

Previously, Masuda et al. reported on the use of AIDAF in digital transformation utilizing design thinking approaches and agile development [9]. In this study, we combine AIDAF with a method for deriving performance indicators. As a result of studying the derivation of performance indicators using AIDAF and Strategic Performance Indicator Derivation Framework (SPIDF), the following four steps were used to define the performance indicators for the application of human-centered agile development.

Table 1. Strategic Performance Indicator Deriving Framework for Design Thinking Approach

	AIDAF for Design Thinking Approach [8]	SPIDF for Design Thinking Approach
Concept Phase	1A. Design Thinking approach for Prototypes in PoC	Performance of Prototypes for User
Design Phase	2A. Design Thinking approach	Performance of Business Model / Business Process
Development Phase	3A. Development of Prototypes into Society	Performance of Prototypes for Business/Society
Operation Phase	4A. Risk Management based on results of AB reviews	Performance of Actual Operation

3.1 Concept Phase: Performance of Prototype for User

In this phase, as a proof of concept for the product, an evaluation of the effect on the user experience is required through prototyping. In order to identify the value to be provided, user needs will be captured, prototyped, and conceptually verified in a design thinking workshop. Performance indicators in this phase will focus on users.

Table 2. SPIDF elements in Concept Phase

Organization State	Analyze what the user needs to accomplish and the pain points in doing so
Organization Enabler	Analyze changes in the user's thoughts, feelings, and behavior required at the point of organization state
Intervention	Set up the functions necessary to solve users' problems and cause changes in their thoughts, feelings, and behavior
Performance Indicator	Based on the objective state and intervention function, performance indicators for the PoC will be designed. As for the evaluation of effectiveness, performance indicators will be derived based on the user's objective state and intervention points to the enabler. In the course of the study, it is possible to change performance indicators in an exploratory manner by reviewing not only the intervention, but also the purpose state and enablers on the user side

3.2   Design Phase: Performance of Business Model/Business Process

In this phase, the business model or business process for product realization is validated through prototyping. Business models and business processes are designed and feasibility is verified. AIDAF supports to evaluate the alignment of the architecture among stakeholders and support the evaluation of feasibility. Performance indicators in this phase can be derived focusing on the business model.

Table 3. SPIDF elements in Design Phase

Organization State	The Business Model Canvas [27] can be used to materialize the objective state
Organization Enabler	Identify enablers that should be in place when the business model is realized
Intervention	Consider initiatives to implement the business model and the business processes
Performance Indicator	Key element of the business model can be used as an initial performance indicator

3.3   Development Phase: Performance of Prototypes for Business/Society

In this phase, the actual system that provides the product or service to the user is established. In addition to alignment with the existing business, performance indicators are required that consider external requirements such as regulations. In AIDAF, the context to be analyzed is broadened to examine business alignment in a broader scope. Performance indicators in this phase should include an assessment that the business is being implemented as planned and that the implemented business is providing the intended value.

Table 4. SPIDF elements in Development Phase

Organization State	Establish the objective state as an initial milestone for the business
Organization Enabler	Identify enablers that should be in place when milestones are achieved
Intervention	Identify the business processes to be implemented to deliver value, as well as the capabilities to address the existing business and external environment
Performance Indicator	Performance indicators related to the outputs from business processes in the business / society may be considered

3.4 Operation Phase: Performance of Actual Operation

This phase involves the operation and maintenance of the established business. In addition to measuring performance indicators to evaluate that the business is continuously achieving its objectives, the performance indicators must be reviewed in light of changes in the external environment. Performance indicators in this phase should include an assessment of whether the business is continuously achieving its objectives.

Table 5. SPIDF elements in Operation Phase

Organization State	Develop the objective state to the next step according to the maturity of the business
Organization Enabler	Identify the enablers that should be in place when each objective state is realized
Intervention	In addition to the implemented features, additional features can be added that should be improved or enhanced
Performance Indicator	It is possible to shift the status of implementation of business processes to outcome evaluation as a business

4   Discussion

4.1 Performance Indicators in Human-Centered Agile Development

It is considered that analyzing the value to users and the business model in correspondence with the SPIDF's objective state and enabler perspectives will support the design of performance indicators related to the outcomes to be aimed for. Also, by revising the analysis each time the intervention is reviewed, it is possible to design performance indicators that correspond to the intervention. Thus, with respect to RQ1, the SPIDF is considered to provide a flexible derivation of performance indicators based on outcomes and interventions.

4.2 Dynamic Management of Performance Indicators in the Process of Digital Transformation

During the proof-of-concept phase regarding value, user-focused performance indicators can be formulated, and during the implementation phase, the focus can be on the business model. When applying the design thinking approach and agile development, the value provided and the business model as the delivery system are reviewed in a short period of time. The performance indicator derivation method allows performance indicators to be reviewed based on the reviewed points. This will be validated by conducting case studies by companies to evaluate changes in performance indicators. Pivoting occurs with high frequency, especially in the early stages of a business. The performance indicator derivation method has traceability between the target state and the intervention when deriving performance indicators, so that modifications can be made based on changes. In addition, external interventions such as regulations and cybersecurity can also be set as performance indicators by extending the perspective of the intervention. In light of the above, with respect to RQ2, the SPIDF is expected to support the revision of performance indicators in accordance with changes in objectives and intervention measures.

4.3 Future Research

We will conduct a case study of this proposal in a pharmaceutical company to verify the derivation of performance indicators with design thinking and agile development.

5 Conclusion

This paper proposes the use of performance indicator derivation methods when applying design thinking and agile development approaches, and suggests that combining the AIDAF framework with the performance indicator derivation methods can support the derivation of performance indicators when performance indicators need to be reviewed in a short period of time. Since this study examines the practical feasibility in a limited context, additional studies expanding the scope of application are considered necessary for generalization. In addition to the intervention by the organization, the impact of external interventions should also be considered on the organization's performance. Therefore, we will also conduct research on how to derive performance indicators that take external influences into account.

References

1. Singh, M., Sachan, S., Singh, A., & Singh, K. K. (2020). Internet of Things in pharma industry: possibilities and challenges, pp. 195–216 Academic Press
2. Sharma, A., Kaur, J., Singh, I.: IoT in pharmaceutical manufacturing, warehousing, and supply chain management. SN Comput. Sci. **1**(4), 1–10 (2020)
3. Szlezak, N., Evers, M., Wang, J., Pérez, L.: The role of big data and advanced analytics in drug discovery, development, and commercialization. Clin. Pharmacol. Ther. **95**(5), 492–495 (2014)
4. Resman, M., Pipan, M., Šimic, M., Herakovič, N.: A new architecture model for smart manufacturing: a performance analysis and comparison with the RAMI 4.0 reference model. Adv. Prod. Eng. Manag, **14**(2), 153–165 (2019)
5. Miura, K., Kobayashi, N., Shirasaka, S.: A strategic performance indicator deriving framework for evaluating organizational change. Rev. Integr. Bus. Econ. Res. **9**(4), 36–46 (2020)
6. Buckl, S., Matthes, F., Schulz, C., Schweda, C.M.: Exemplifying a Framework for Interrelating Enterprise Architecture Concerns. In: Sicilia, M.-A., Kop, C., Sartori, F. (eds.) ONTOSE 2010. LNBIP, vol. 62, pp. 33–46. Springer, Heidelberg (2010). https://doi.org/10.1007/978-3-642-16496-5_3
7. Tamm, T., Seddon, P.B., Shanks, G., Reynolds, P.: How does enterprise architecture add value to organizations? Commun. Assoc. Inform Syst. **28**(1), 141–168 (2011).
8. Masuda, Y., Shirasaka, S., Yamamoto, S., Hardjono, A.B., Practices in Adaptive Enterprise Architecture with Digital Platform: A Case of Global Healthcare Enterprise, T.: International Journal of Enterprise Information Systems. IGI Global. **14**, 1 (2018)

9. Masuda, Y., Zimmerman, A., Shepard, D. S., Schmidt, R., Shirasaka, S.: An adaptive enterprise architecture design for a digital healthcare platform: toward digitized society-industry 4.0, Society 5.0. In: 2021 IEEE 25th International Enterprise Distributed Object Computing Workshop (EDOCW), (pp. 138–146) (2021). IEEE

10. Dixon, J.R., Nanni, A.J., Vollmann, T.E.: New performance challenge: Measuring operations for world-class competition. McGraw-Hill Professional Publishing, Homewood (1990)

11. Kaplan, R.S., Norton, D.P.: Linking the balanced scorecard to strategy. Calif. Manage. Rev. **39**(1), 53–79 (1996)

12. Bourne, M., Melnyk, S., Bititci, U.S.: Performance measurement and management: theory and practice. Int. J. Oper. Prod. Manag. **38**(11), 2010–2021 (2018)

13. Bititci, U., Garengo, P., Dörfler, V., Nudurupati, S.: Performance measurement: challenges for tomorrow. Int. J. Manag. Rev. **14**(3), 305–327 (2012)

14. Rossi, P.H., Lipsey, M.W., Henry, G.T.: Evaluation: A systematic approach. Sage publications (2018).

15. International Organization for standardization (2008), ISO9001:2008 Quality management systems – requirements

16. ICH. ICH Harmonized Tripartite Guideline: Pharmaceutical Quality System Q10; June, 2008

17. Lawrence, X.Y., Kopcha, M.: The future of pharmaceutical quality and the path to get there. Int. J. Pharm. **528**(1–2), 354–359 (2017)

18. Arling, E.R., Dowling, M.E., Frankel, P.A.: Creating and managing a quality management system. Pharmaceutical Sciences Encyclopedia: Drug Discovery, Development, and Manufacturing, pp. 1–48 (2010)

19. Iizuka, Y., Nagata, Y.: Introduction to Modern Quality Management (Japaneses). Asakura Publishing Co., Ltd (2009)

20. Curedale, R.: Design Thinking Processes & Methods. Design Community College, Los Angeles, (2018)

21. Stanford d.school bootcamp bootleg, https://dschool.stanford.edu/resources/the-bootcamp-bootleg

22. Peterson, K., Wohlin, C., Baca, D.: The waterfall model in large-scale development, in International Conference on Product-Focused Software Process Improvement, 2009, S. 386–400

23. Fowler, M., & Highsmith, J., The agile manifesto, Software development, Bd. 9, Nr. 28–35, 2001

24. Qurashi, S.A., Qureshi, M.: Scrum of scrum solution for large size teams using scrum methodology, arXiv preprint arXiv: 1408.6142 (2014)

25. Leffingwell, SAFe 4.5 reference guide: scaled agile framework for lean enterprises. Addison-Wesley Professional, 2018

26. Ambler, S.W., Lines, M.: Disciplined agile delivery: A practitioner's guide to agile software delivery in the enterprise. IBM press, 2012

27. Osterwalder, A., Pigneur, Y.: Business Model Generation. John Wiley, Hoboken, NJ, USA (2010)

Human-Centred Design Thinking Using the Intelligence Amplification Design Canvas and the Adaptive Integrated Digital Architecture Framework

Jean Paul Sebastian Piest[1]([✉]) [ID], Yoshimasa Masuda[2,3] [ID], Osamu Nakamura[2], and Koray Karaca[1] [ID]

[1] University of Twente, Drienerlolaan 5, 7522 NB Enschede, The Netherlands
j.p.s.piest@utwente.nl
[2] Keio University, Minato City, 2 Chome-15-45 Mita, Minato City, Tokyo 108-8345, Japan
[3] Carnegie Mellon University, 5000 Forbes Avenue, Pittsburgh, PA 15213, USA

Abstract. Human-Centred Design (HCD) is essential to realize the Society 5.0 vision of the supersmart information society. Design thinking is a popular approach which has been incorporated in the Adaptive Integrated Digital Architecture Framework (AIDAF). However, design thinking is less systematic when compared to the ISO 9241-210:2019 for the HCD of interactive systems. Extending earlier work, this paper aims to explore how the ISO 9241-210:2019 for HCD of interactive systems and Intelligence Amplification (IA) design canvas can be incorporated in the AIDAF for design thinking approach. This study utilized the ISO 9241-210:2019 for HCD of interactive systems as an evaluation framework in combination with exploratory research. First, this study described how the six principles and four design activities of the ISO 9241-210:2019 for HCD of interactive systems can be applied as part of the AIDAF for design thinking approach. Second, this study explored how the IA design canvas can be incorporated in the AIDAF for design thinking approach for prototyping and enterprise systems development. Third, a set of seventeen testable propositions were defined for future evaluation. The main limitations of this study are the partial use of the ISO 9241-210:2019 for HCD of interactive systems, the focus on the AIDAF in healthcare using agile development methodologies, and its exploratory nature. Current work focuses on case study research and in-depth evaluation using the ISO 9241-210:2019 for HCD of interactive systems. Future work may contribute by conducting a comparison study of alternative or hybrid development methodologies.

Keywords: Human-Centred Design · ISO 9241-210:2019 · Design Thinking · AIDAF · Intelligence Amplification · Design Canvas

1 Introduction

Understanding the needs and role of people in the context in which interactive systems are used is essential for Human-Centred Design (hereafter HCD) [1]. HCD emerged from consumer and function-oriented design and developed into a humanistic design approach

© The Author(s), under exclusive license to Springer Nature Singapore Pte Ltd. 2023
A. Zimmermann et al. (Eds.): KES-HCIS 2023, SIST 359, pp. 153–163, 2023.
https://doi.org/10.1007/978-981-99-3424-9_15

[1]. In the past decades, HCD research resulted in an extensive body of knowledge [1] and a large set of tools, methods, and techniques [2]. The ISO 9241-210:2019 defines HCD as "an approach to interactive systems development that aims to make systems usable and useful by focusing on the users, their needs and requirements, and by applying human factors/ergonomics, and usability knowledge and techniques" [3]. This standardized and systematic design approach is based on more than ten years of application and replaced ISO 9241-210:2010 and ISO 13407 [3].

Design Thinking (hereafter DT) is a popular design approach that includes aspects of HCD, but tends to have a less systematic approach when compared to the ISO 9241-210:2019 for HCD of interactive systems. DT is originally introduced by Tim Brown, became well-known and widespread via the design school and innovation management research at the Stanford University, and evolved to a prominent design approach in past decade [4]. Related work urges the need for empirical research regarding the practices of creating and sustaining a design culture in organizations where people from different disciplines collaboratively create concrete and valuable solutions for real-life challenges [4]. In response to this need, this paper will connect current research and development work of the OHP2030 consortium and the Adaptive Integrated Digital Architecture Framework (AIDAF) for DT approach [5, 6].

Earlier work described how DT can be incorporated as part of the AIDAF in a healthcare community [5]. The AIDAF for DT approach goes beyond the scope of the five-step DT process by implementing the prototype in society or a production environment. It is important to conduct a risk assessment prior implementation, especially in the high-stake context of healthcare. Therefore, the DT approach is extended with the STrategic Risk Mitigation Model (STRMM) to assess risks. Currently, the AIDAF for DT approach is used with the Open Healthcare Platform 2030 (OHP2030) consortium [6]. Other earlier work introduced the Intelligence Amplification (IA) design canvas [7], which incorporates principles of HCD, and related design workshop approach [8]. The present study connects the AIDAF for DT approach to the IA design canvas.

Extending earlier work, the aim of this paper is to explore how the ISO 9241-210:2019 for HCD of interactive systems and IA design canvas can be incorporated in the AIDAF for DT approach. The following Research Questions (RQs) were defined:

- **RQ1:** How can the ISO 9241-210:2019 for HCD of interactive systems be incorporated within the AIDAF for DT approach?
- **RQ2:** How can the IA design canvas support prototyping and enterprise system development as part of the AIDAF for DT approach?

This study utilizes the ISO 9241-210:2019 for HCD of interactive systems as an evaluation framework and investigates how the IA design canvas can be incorporated within the AIDAF for DT approach using exploratory research. More specifically, this research focuses on defining testable propositions for DT workshops, prototyping, and case study research in the OHP2030 consortium in preparation for future evaluation.

This paper is structured as follows. Section 2 discusses related and earlier work. Section 3 explains the methodology. Section 4 presents the HCDT approach with the AIDAF and IA design canvas and testable propositions for future evaluation. Section 5 concludes the study and positions future work.

2 Related and Earlier Work

This section discusses related and earlier work. Section 2.1 reviews the main parts of the ISO 9241-210:2019 for HCD. Section 2.2 summarizes earlier work.

2.1 ISO 9241-210:2019 for HCD

The ISO 9241-210:2019 for HCD of interactive systems is built around user needs and unifies the context of use and the needs and requirements of users with processes to produce and evaluate design solutions [3]. The ISO 9241-210:2019 for HCD of interactive systems contains six principles: 1) design based on explicit understanding of users, tasks, and environments, 2) user involvement throughout the design and development processes, 3) user-centred evaluation to drive design, 4) iterative processes and refinement, 5) design that encompasses the full experience, and 6) multidisciplinary design team and perspectives [3]. Consequently, four interdependent HCD design activities are defined: 1) understanding context of use, 2) specifying user requirements, 3) producing design solutions, 4) evaluating the design [3]. From a practical perspective, the ISO 9241-210:2019 for HCD of interactive systems provides guidelines to produce a plan for HCD of interactive systems and detailed process descriptions. Furthermore, the evaluation and conformance are supported by assessment instruments and checklists. Additional to the HCD processes, standard processes have been defined and described for enabling, executing, and assessing HCD within organizations. Here, the adoption starts with commitment to the six principles and four design activities, but the standard covers the full life cycle. The above is related to the need for empirical research and current work in the OHP2030 consortium for the purpose of this study.

2.2 Earlier Work

The AIDAF for DT approach was initially positioned and applied for the design and development a physical activity monitoring system [5]. Here, a design approach is presented based on DT and agile software development consisting of four steps. More specifically, the AIDAF was applied to design a digital healthcare platform taking into account the needs and cultural preferences of users. Additionally, the guidelines of SAP Fiori were used for user interface design and privacy and safety aspects were taken into account. The demonstrated approach illustrates how users were involved throughout the design and development. Furthermore, it shows how risks were assessed using the SRMM. Implicitly, the above can be related to the six principles and four design activities of the ISO 9241-210:2019 for HCD of interactive systems. This study will refine the AIDAF for DT approach by explicitly incorporating the principles and design activities in preparation for future development work in the OHP2030 consortium.

The IA design canvas is designed and developed as part of an industry platform for data-driven logistics in small and medium-sized enterprises [9]. The IA design canvas is tested and applied in various projects and different settings using action design research [7]. The IA design canvas is based on HCD and aims to ease design processes and improve communication between users and experts. The design workshop contributes to understanding stakeholder needs and user requirements in relation to an artefact in context and is based on iterative processes involving a multidisciplinary team [8]. The present study investigates how the IA design canvas can be utilized throughout the full lifecycle by incorporating its use in the AIDAF for DT approach.

3 Methodology

This section explains the research methodology. Section 3.1 discusses the use of the ISO 9241-210:2019 for HCD in this study. Section 3.2 discusses how the explorative study was conducted.

3.1 Application of the ISO 9241-210:2019 for HCD

This study was conducted using established design approaches, as introduced in Sect. 1 and described in Sect. 2.1. This section will explain the application of the ISO 9241-210:2019 for HCD of interactive systems.

The ISO 9241-210:2019 for HCD of interactive systems was selected to refine the AIDAF for DT approach for the design and development of healthcare applications in the OHP2030 consortium and alignment with the Society 5.0 vision of the "human-centred supersmart society" [10]. The rationale, described in the ISO 9241-210:2019 for HCD of interactive systems [3], emphasizes that highly usable systems provide substantial benefits for their users, employers, and involved stakeholders, and tend to be more successful from both a technological and commercial perspective.

The six principles and four design activities, described in Sect. 2.1., were used to refine the AIDAF for DT approach for prototyping and enterprise system development. The expected benefits were formulated in the form of testable propositions that can be used for future evaluation. Based on the outcomes of this study, the adoption of the ISO 9241-210:2019 for HCD of interactive systems will be considered in more detail. This study does not intend to evaluate the AIDAF for DT approach for certification.

3.2 Exploratory Research

Following the aim of this study, an explorative study has been designed and conducted, because this study involves the first-time use of the IA design canvas as part of the AIDAF for DT approach. The initial scope was set to healthcare to extend earlier work and prepare for future evaluation in a case study as part of the OHP2030 consortium.

The refined approach was developed by the main ideators of both the IA design canvas and the AIDAF (namely, the first and second author). Then, a group of three experts (including the third author) was formed to verify the refined approach. These experts represented domain knowledge and experience with software development projects in healthcare (medical doctor) and expertise regarding system design, engineering, and management (professor and doctor in computer science). The first and second author presented the refined approach with testable propositions for DT and prototyping and enterprise systems development. Additionally, the first author demonstrated the use of the IA design canvas using an example in healthcare. Next, the proposed refined approach and use of the IA design canvas were discussed. After the verification by the three experts, the authors revisited the refined DT approach and testable propositions for prototyping and enterprise systems development.

Current work in progress aims to evaluate the use of the IA design canvas for the AIDAF DT approach by organizing IA design canvas workshops, developing, and implementing a prototype, and preparing case study research to evaluate the use of the IA design canvas for enterprise systems development. Therefore, testable propositions have been defined for both prototyping and enterprise software development in support of these future evaluation.

Now that the methodology has been explained, the next section presents the HCDT approach with the AIDAF and IA design canvas.

4   HCDT Approach with the AIDAF and IA Design Canvas

This section presents the HCDT approach with the AIDAF and IA design canvas. Section 4.1 positions DT as part of the AIDAF to answer RQ1. Section 4.2 discusses the use of the IA design canvas to answer RQ2. Section 4.3 contains testable propositions for DT and prototyping. Section 4.4 includes testable propositions for enterprise software development.

4.1   Positioning HCD and DT as Part of the AIDAF

The AIDAF is a strategic Enterprise Architecture Framework (EAF) that enables digital agility based on an Adaptive Enterprise Architecture (AEA) cycle, as shown in Fig. 1.

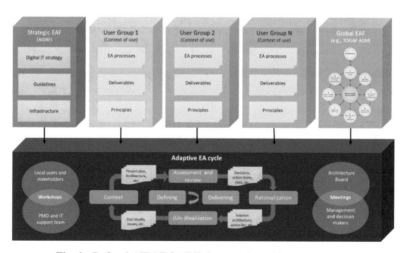

Fig. 1.   Refined AIDAF for DT (image adapted based on: [11]).

The AEA of the AIDAF was enhanced with workshops and meetings, as shown in Fig. 1, based on the six principles and four design activities of the ISO 9241-210:2019 for HCD of interactive systems. Workshops with multidisciplinary teams focus on identifying and understanding stakeholder needs, user groups, and user requirements in the context of use taking into account the full lifecycle (including (un-)realization). The user-oriented and iterative approach of agile software development facilitates continuous user involvement, verification, and refinement of work products, and user-centred evaluation throughout the full lifecycle. The IA design canvas can facilitate the initial definition and documentation of the context of use, stakeholder needs, user requirements, goals, and metrics in a compact and comprehensible manner. The completed IA design canvas can serve as the starting point to develop detailed specifications, mock-ups of the user interface and interaction, and low- or high-fidelity prototypes for user group(s) in a specific context of use, as shown in Fig. 2. Additionally, the scope of DT may be extended to enterprise systems development using agile development (--- line).

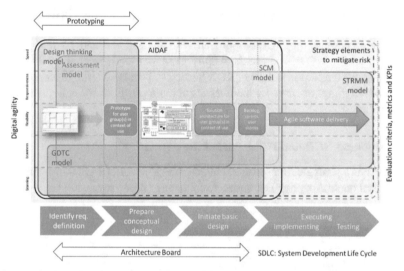

Fig. 2. Positioning of the IA design canvas for DT and prototyping and envisioned full life cycle coverage (--- line) as part of the AIDAF and related models (image adapted based on: [11]).

4.2   AIDAF for DT Approach with the IA Design Canvas

Following the previous section, Table 1 presents an overview of the refined AIDAF for DT approach and proposed use of the IA design canvas.

4.3   Testable Propositions for DT and Prototyping

This section presents and discusses the nine testable propositions that were defined for DT and prototyping, as presented in Fig. 3.

In step 1A of the AIDAF for DT approach, the IA design canvas may improve the communication between local user groups and experts by collaboratively emphasizing and defining ideas and the context of use. Additionally, the IA design canvas can facilitate iterative requirements gathering and initial verification by local partners. As a result of

Table 1. Overview of refined AIDAF for DT approach and use of the IA design canvas.

Stage	AIDAF for DT approach	Use of the IA design canvas
Emphasize	"[1A] DT approach with partners in a healthcare community is utilized in designing user interfaces with cultural preferences, before implementing prototypes for Digital Platforms" [5]	[1A-1.1] Project manager: Plan the project, form a group with different local users, IT experts, and a facilitator [1A-1.2] Participants: Complete preparation steps (e.g., form duos, brainstorm for ideas, and select idea for workshop) [1A-1.3] Facilitator: Conduct (online) intake prior the workshop
Define		[1A-2.1] Facilitator: introduction of the workshop and participants [1A-2.2] Participants: Complete Step 1 in the workshop: define idea and context of use in 2–3 sentences [1A-2.3] Participants: pitch idea and context of use to fellow participants [1A-2.4] Participants: complete Step 2–5 in workshop using the 4 principles to systematically define the 13 elements [1A-2.5] Participants: Pitch the final IA design canvas to fellow participants [1.A-2.6] Participants: Refine the IA design canvas based on feedback
Ideate		[1A-3.1] Participants: craft one solution using IA design canvas [1A-3.2] Participants: ideate solution alternatives for elements within the IA design canvas or craft and compare multiple IA design canvases

(continued)

Table 1. (*continued*)

Stage	AIDAF for DT approach	Use of the IA design canvas
Prototype	"[1A] Agile software development is utilized to develop a prototype" [5]	This step is not in scope of the IA design canvas. However, the back of the IA design canvas can be used to draw a conceptual solution architecture or sketch the envisioned UI/UX design
Test	"[1A] The prototype is tested and refined with architecture guidelines (e.g., user interfaces and privacy) as part of architecture reviews" [5]	This step is not in scope of the IA design canvas approach. However, the goals and metrics (and other relevant) elements can guide testing and validation of the prototype
Design for production	"[2A] In Context phase of the AEA cycle in the AIDAF, the project manager can adopt the DT approach for enterprise systems development to define necessary enhancements of the prototypes" [5]	This step is not in scope of the IA design canvas approach. However, the systems and integration (and other relevant) elements can provide input for the initial scope
Implementation of prototype in society	"[3A] In Assessment/Architecture Review phase in the AIDAF, the digital IT project's proposal with enhancements of prototypes can be reviewed for deployment into society and in consideration of production environments, rationalization and (un-) realization of systems" [5]	This step is not in scope of the IA design canvas approach. However, the IA design canvas provides 13 elements that can be incorporated and further described in a project plan/proposal
Risk management	"[4A] In the (un-)realization phase in the AIDAF, risk management process can be started based on the review's results and necessary policies (i.e. privacy). In digital IT projects, project managers can cope with risks using strategy elements for risk mitigation from the architecture board" [5]	This step is not in scope of the IA design canvas approach. However, social-ethical-legal aspects can be identified in an early stage

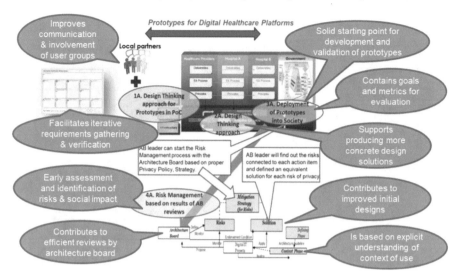

Fig. 3. Overview of testable propositions for the use of the IA design canvas for DT and prototyping (image adapted based on: [5]).

step 1A, the use of the IA design canvas is expected to result in more concrete design solutions. This may form a solid starting point for the development and validation of prototypes in step 3A. Here, the goals and metrics may be used for evaluation. Following, the IA design canvas may contribute to early assessment and identification of risks as part of step 4A. Furthermore, the IA design canvas can contribute to explicit understanding of the context of use and improving initial designs. Lastly, the IA design canvas may contribute to more efficient reviews by the architecture board.

4.4 Testable Propositions for Enterprise System Development

This section highlights and briefly discusses the eight testable propositions that were defined for enterprise system development, as shown in Fig. 4.

In the context of enterprise systems development, the use of the IA design canvas may contribute to digital agility. In relation to the assessment model, the IA design canvas may result in a more structured approach to assess project proposals. More specifically, the IA design canvas is expected to support the process of defining user requirements and/or -stories and verification of software designs. Extending the use during design workshops, the IA design canvas could be utilized for continuous involvement of user groups and stakeholders and improving communication. As a result of the previous, the IA design canvas might result in valuable feedback to improve/refine the solution architecture. Additionally to initial risk assessment, the IA design canvas may contribute to contextual risk assessment with explicit context of use descriptions. Lastly, as a result of the previous, the IA design canvas might contribute to better progress monitoring and evaluation during the software development life cycle.

Next, this study will be concluded and future work will be positioned.

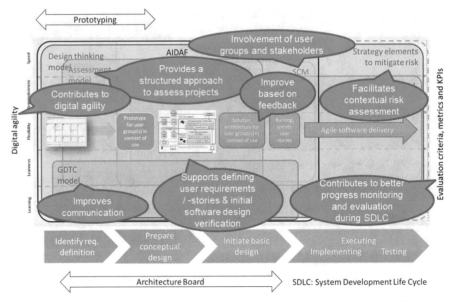

Fig. 4. Overview of testable propositions for the use of the IA design canvas for enterprise system development (image adapted based on: [11]).

5   Conclusion

This paper presented how the ISO 9241-210:2019 for HCD of interactive systems and the IA design canvas can be incorporated in the AIDAF for DT approach for prototyping and enterprise systems development. First, the DT approach has been positioned as part of the AIDAF. Next, the use of the six principles and four design activities of the ISO 9241-210:2019 for HCD of interactive systems and the IA design canvas have been described. Then, the current AIDAF for DT approach was refined and the use of the IA design canvas was tabulated and described. Following, nine testable propositions for DT and prototyping and eight testable propositions for enterprise software development were defined for future evaluation in case study research.

The main limitation of this study is the scope of the explorative study and focus on the AIDAF for DT approach and agile software development in the context of healthcare. Another limitation is the verification by three experts. The refined approach has not yet been evaluated in a case study. Moreover, the application of the ISO 9241-210:2019 for HCD of interactive systems is limited to the six principles and four design activities. Evaluation and certification were not part of this study.

Current work focuses on conducting case study research and in-depth evaluation using the ISO 9241-210:2019 for HCD of interactive systems. Additionally, the use of the business model canvas and value proposition canvas are being explored.

Future work may contribute by exploring the use of alternative or hybrid development approaches (e.g., waterfall, systems engineering, scaled agile, agile-V) to evaluate the testable propositions and compare the results with agile software development.

References

1. Zhang, T., Dong, H.: Human-centred design: an emergent conceptual model. http://bura.bru nel.ac.uk/handle/2438/3472 (2009).
2. Giacomin, J.: What Is Human Centred Design? Des. J. **17**(4), 606–623 (2014). https://doi.org/10.2752/175630614X14056185480186
3. ISO (2019). Ergonomics of human-system interaction — Part 210: Human-centred design for interactive systems (ISO 9241–210:2019). URL (Accessed 29 Jan 2023 https://www.iso.org/obp/ui/#iso:std:iso:9241:-210:ed-2:v1:en
4. Auernhammer, J., Roth, B.: The origin and evolution of Stanford University's design thinking: From product design to design thinking in innovation management. J. of Product Innovation Management **38**(6), 623–644 (2021). https://doi.org/10.1111/jpim.12594
5. Masuda, Y., Zimmermann, A., Shepard, D.S., Schmidt, R. Shirasaka, S.: An Adaptive Enterprise Architecture Design for a Digital Healthcare Platform : Toward Digitized Society – Industry 4.0, Society 5.0. 2021 IEEE 25th International EDOCW, 2021, pp. 138–146 https://doi.org/10.1109/EDOCW52865.2021.00043
6. Masuda, Y., Zimmermann, A., Sandkuhl, K., Schmidt, R., Nakamura, O., Toma, T.: Applying AIDAF for Enabling Industry 4.0 in Open Healthcare Platform 2030. In: Zimmermann, A., Howlett, R.J., Jain, L.C., Schmidt, R. (eds.) KES-HCIS 2021. SIST, vol. 244, pp. 211–221. Springer, Singapore (2021). https://doi.org/10.1007/978-981-16-3264-8_20
7. Piest, J.P.S., Iacob, M.E., & Wouterse, M.J.T. (2022). Designing Intelligence Amplification: a Design Canvas for Practitioners. In: AHFE ,vol. 68, pp. 68-76). [8].https://doi.org/10.54941/ahfe1002714
8. Piest, J.P.S., Iacob, M.E., & Wouterse, M.J.T. (2022). Designing Intelligence Amplification: Organizing a Design Canvas Workshop. In AHFE (Vol. 68, p. 247 251). [33]. https://doi.org/10.54941/ahfe1002739
9. Piest, J.P.S.: An Industry Platform for Data-driven Logistics in Small and Medium-sized Enterprises. UT. doi **10**(3990/1), 9789036553650 (2022). https://doi.org/10.3990/1.978903 6553650
10. Cabinet Office. (2023). Society 5.0. https://www8.cao.go.jp/cstp/english/society5_0/index.html [Accessed: 03 Feb 2023]
11. Masuda, Y., Viswanathan, M.: Enterprise architecture for global companies in a digital it era: adaptive integrated digital architecture framework (AIDAF). Springer (2019). https://doi.org/10.1007/978-981-13-1083-6

Applying AIDAF for Digital Transformation Toward Ecosystem in Global Enterprise

Yoshimasa Masuda[1,2,6,8(✉)], Rashmi Jain[3], Alfred Zimmermann[4], Rainer Schmidt[5], Osamu Nakamura[6], and Tetsuya Toma[7]

[1] Carnegie Mellon University, Pittsburgh, USA
ykmasuda@gmail.com
[2] Tokyo University of Science, Keio University, Tokyo, Japan
[3] Computer Science, Monclair State University, New Jersey, NJ, USA
[4] Computer Science, Reutlingen University, Reutlingen, Germany
[5] Munich University of Applied Sciences, Munich, Germany
[6] Graduate School of Media and Governance, Keio University, Kanagawa, Japan
[7] Graduate School of System Design and Management, Keio University, Kanagawa, Japan
[8] Institute of NTT Data Management Consulting, Inc. Tokyo, Tokyo, Japan

Abstract. Enterprises and societies currently face essential challenges, and digital transformation can contribute to their resolution. Enterprise architecture (EA) is useful for promoting digital transformation in global companies and information societies covering ecosystem partners. The advancement of new business models can be promoted with digital platforms and architectures for Industry 4.0 and Society 5.0. Therefore, products from the sector of healthcare, manufacturing and energy, etc. can increase in value. The adaptive integrated digital architecture framework (AIDAF) for Industry 4.0 and the design thinking approach is expected to promote and implement the digital platforms and digital products for healthcare, manufacturing and energy communities more efficiently. In this paper, we propose various cases of digital transformation where digital platforms and products are designed and evaluated for digital IT, digital manufacturing and digital healthcare with Industry 4.0 and Society 5.0. The vision of AIDAF applications to perform digital transformation in global companies is explained and referenced, extended toward the digitalized ecosystems such as Society 5.0 and Industry 4.0.

Keywords: Digital Transformation · Enterprise Architecture · Digital IT · Digital Healthcare · Industry 4.0 · Society 5.0

1 Introduction

Continuous changes are hallmarks of many global companies and information societies, such as the development of new technologies, globalization, shifts in customer needs and new business models. Recently, digital transformation has brought great changes to existing enterprises, ecosystems and economies [5]. Significant changes in cutting-edge IT technology due to recent developments in Cloud Computing and Mobile IT (such as

© The Author(s), under exclusive license to Springer Nature Singapore Pte Ltd. 2023
A. Zimmermann et al. (Eds.): KES-HCIS 2023, SIST 359, pp. 164–176, 2023.
https://doi.org/10.1007/978-981-99-3424-9_16

progress in Big Data technology) have emerged as new trends in information technology. Furthermore, major advances in these technologies and processes have created a "digital IT economy," bringing about business opportunities along with business risks, and forcing enterprises to innovate or face the consequences [7]. Enterprise Architecture (EA) usefully contributes to the design of large integrated systems, helping to address a major technical challenge toward the era of Cloud, Mobile IT, Big Data, and Digital IT in digital transformation. From a comprehensive perspective, EA encompasses all enterprise artifacts, such as businesses, organizations, applications, data, and infrastructure, to establish the current architecture visibility and future architecture/roadmap. On the other hand, EA frameworks need to embrace change in ways that consider the emerging new paradigms and requirements affecting EA, such as mobile IT and the cloud [8, 9].

In the healthcare and manufacturing industries today, new enhancements to business structure and process efficiency through digital platforms such as portals and social networking services (SNSs) are being considered by companies and corporations. Industry 4.0 offers many opportunities for companies to increase flexibility and efficiency in production processes, enabling new business models through Industry 4.0 digital platforms [1, 4]. Society 5.0 can contribute to a supersmart society covering healthcare industries [43].

In light of these developments, a previous study proposed the "Adaptive Integrated EA Framework" to align with the IT strategy to promote Cloud, Mobile IT and Digital Platform, and verified this in the case study [10]. This EA framework was named as the "Adaptive Integrated Digital Architecture Framework - AIDAF" [11].

2   Related Works

2.1   Digital IT and EA for Digital Healthcare, Manufacturing, Smart Energy

In the past decade, EA has become an important method for modeling the correlation for overall images of corporate and individual systems. In ISO/IEC/IEEE42010:2011, architecture framework is defined as "principles, and practices for the architecture descriptions established within a specific domain of application and/or community." Furthermore, EA visualizes the current corporate IT/business landscape to promote a desirable future IT model [9]. It is not a simple support activity [8], and it offers many benefits to companies, such as coordination, communication, and planning between business and IT, and reduction in the complexity of IT [35]. To deliver these benefits, EA frameworks need to cope with the emerging new paradigms such as Cloud computing or enterprise mobility [8].

Mobile IT computing is an emerging concept using Cloud services provided over mobile devices [40]. In addition, Mobile IT applications are composed of Web services. Many studies discuss the integration of EA with Service Oriented Architecture (SOA), except for Mobile IT. The SOA architecture pattern defines the four basic forms of business service, enterprise service, application service, and infrastructure service [39]. The OASIS, which is a public standards group [37], introduces an SOA reference model. Many organizations have invested in SOA as an approach to manage rapid change [36]. Meanwhile, attention has been focused on Microservices architecture, which allows rapid adoption of new technologies, such as Mobile IT, Cloud computing and platforms

[38]. SOA and Microservice vary greatly from service characteristics perspective [39]. Microservice is an approach for dispersed systems that is defined by the two basic forms of functional services through an API layer and infrastructure services. Multiple Microservices cooperating to work together enable the implementation as a Mobile IT application [6].

For cloud computing, the NIST defined three cloud service models such as software as a service (SaaS), platform as a service (PaaS), and infrastructure as a service (IaaS) [13]. PaaS is an IaaS platform that includes both system software and an integrated development environment. SaaS is a software application developed, implemented, and operated on a PaaS foundation. IaaS accommodates PaaS and SaaS by offering infrastructure resources, such as computing network storage memory through specific centers [13]. Cloud computing is a cost-effective option for acquiring strong computing resources to deal with big data, with significant adoption in the healthcare industry [12]. Many Mobile IT applications also operate with SaaS Cloud-based software [40]. The integration and relationship between EA and Cloud computing are discussed rarely in literature. Considering the recent dynamic moves in Cloud computing, companies must link the service characteristics of EA and Cloud computing [17]. The traditional approach takes months to develop an EA realizing a Cloud adoption strategy, and organizations will demand adaptive enterprise architecture to iteratively develop and manage an EA adaptive to the Cloud technology [41]. The implementation of Big Data analytics in healthcare is advancing, enabling the exploration of large data sets incorporating electronic healthcare records (EHRs) to uncover hidden patterns, unknown correlations, and other useful information [14, 15, 31]. Advances in Big Data analytics can help transform research situations from descriptive to predictive and prescriptive [16].

The term "Internet of Things (IoT)" refers to "the collection of uniquely identifiable objects embedded in or accessible through Internet hosts" [7], such as interaction devices, smart homes, other smart life scenarios. The current state of research for the Internet of Things architecture [18] lacks a holistic understanding of EA and management [19–21], showing a range of physical standards, methods, tools, and a large number of heterogeneous IoT devices [22]. Zimmermann et al. proposed a first reference architecture (RA) for the IoT [22] in context of digital enterprise architectures.

IoT can be the main enabler for distributed healthcare applications [23], therefore, potentially can contribute to the overall decrease of healthcare costs while increasing health outcomes, although behavioural changes of the stakeholders are required [16, 23]. Internet of Things (IoT) can change the face of robotics by proposing next generation class of intelligent robotics titled as "Internet of Robotic Things (IoRT)," in the near future in collaboration with artificial intelligence, machine learning, deep learning and cloud computing [27]. The Internet of Medical Robotics Things (IoMRT) is playing a crucial role in medical environments to enhance the effectiveness of using medical devices, speed, and operating accuracy. The IoMRT can be utilized to collect the patients' health data with sensors and devices connected to the internet-based health monitoring systems through online networks [3].

Moreover, according to previous research [42], when promoting Cloud platforms, Mobile IT, Big Data and IoT solutions strategically, it is proposed as a good option that a company that applies The Open Group Architecture Framework (TOGAF) or Federal

Enterprise Architecture Framework (FEAF) can adopt the integrated framework with the Adaptive EA framework supporting elements of Cloud computing.

2.2 Industry 4.0 and Society 5.0

The digitization of global industries and value chains and the associated need for structured research and standardization has given rise to four major initiatives in the USA, China, Japan and Germany. These initiatives address potentials and challenges of digitalization [1]. Industry 4.0 is dedicated to research for German industry and supports the implementation of this vision in manufacturing companies. The Industry 4.0 platform identified 17 technology development fields in its "Recommendations for the implementation of the Industry 4.0 strategic initiative", covering essential aspects of Industry 4.0 and a roadmap [1, 4, 28].

According to Japanese government documents, Society 5.0 can be defined "through the high degree of fusion between cyberspace and physical space, economic progress can be aligned with solving social problems by providing goods and services to meet repeated latent needs regardless of location, age, gender, or language" [43].

As Table 1 of [43] shows, there are some commonalities between Industry 4.0 and Society 5.0. Both visions emphasize the use of technology, including IoT-related technology, AI, and Big Data analysis [43]. There are some differences, however. Industry 4.0 advocates smart factories, while Society 5.0 calls for a supersmart society. In terms of the future technological innovations, Industry 4.0 calls for an industrial revolution centered on manufacturing, whereas, Society 5.0 focuses heavily on the public impact of technology [43].

2.3 AIDAF Framework

Over the past decade, EA has become an important framework for modeling the relationship between enterprise and individual systems. In ISO/IEC/ IEEE42010:2011, an architecture framework is defined as "conventions, principles, and practices for the description of architecture established within a specific domain of application and/or community of stakeholders" [34]. EA is an essential element of corporate IT planning and offers benefits to companies, like coordination between business and IT [35].

Chen et al. have discussed the integration of EA with service-oriented architecture (SOA) [36]. OASIS, a public standards group [37], introduces an SOA reference model. Meanwhile, attention has been focused on microservice architecture, which allows rapid adoption of new technologies like Mobile IT, IoT and cloud computing [38]. SOA and Microservice vary greatly from the viewpoint of service characteristics [39]. Microservice is an approach for dispersed systems defined from the two basic forms of functional services through an application programming interface (API) layer and infrastructure services [38].

In terms of Cloud Computing, many Mobile IT applications operate with SaaS Cloud-based software [40]. Traditional EA approaches require months to develop an EA to achieve a Cloud adoption strategy, and organizations will demand adaptive EA to iteratively develop and manage an EA for Cloud technologies [41]. Moreover, few studies discussed EA integration with Mobile IT [11]. From the standpoint of EA for cloud computing, there should be only an adaptive EA framework that is supporting elements of cloud computing [42]. Moreover, according to the previous survey research [42], when promoting Cloud/Mobile IT in a strategic manner, a company that has applied TOGAF or FEAF can adopt the integrated framework using the adaptive EA framework supporting elements of Cloud Computing.

A preliminary research of this paper proposed an Adaptive Integrated EA framework depicted in Fig. 1 of this preliminary research paper, which should integrate with IT strategy promoting Cloud, Mobile IT, Digital IT, and evaluated this in the case study [10]. In the adaptive EA cycle, project plan documents including architecture for new digital IT projects should be made on a short-term basis in the context phase by referring to materials of the defining phase (e.g., architectural guidelines aligned with IT strategy) per business needs. During the Assessment/Architecture Review Phase, the architecture board (AB) reviews the architecture in the initiation documents for the IT project. In the Rationalization Phase, the stakeholders and AB decide upon replaced or decommissioned systems by the proposed new information systems. In the Realization Phase, the project team starts implementing the new IT project after weighing issues and action items [10, 11]. In the adaptive EA cycle, organizations can deploy an EA framework such as TOGAF and a simple EA framework based on an operational domain unit in the upper part of Fig. 1 of [10, 11]. Furthermore, the fast step architecture out (FSAO) process for digital transformation with AIDAF was proposed and verified [24], and the adaptive integrated architecture board framework (AIABF) was verified to highlight suitable functions of digital platforms and necessary deliverables for digital transformation in the AIDAF [24].

3   Digital Transformation Process of Adaptive Enterprise Architecture – FSAO Approach

As a result of investigating the tasks and steps for digital transformation with the Architecture and project management office (PMO) organizations of global companies, the author proposed and verified the "FSAO process for digital transformation with AIDAF" based on the tasks and descriptions to facilitate the development of digital EA deliverables and the efficient and faster processing of digital transformation, in the case studies of global companies covering global healthcare enterprise (GHE) and global manufacturing company (GMC) [24], as shown in Fig. 1.

The Reference Architectural Model Industry 4.0 (RAMI 4.0) was developed by the Platform 4.0 in 2015. It consists of several layers, hierarchical levels, and the product lifecycle representing the value stream [4]. By populating RAMI 4.0 with generic technologies while applying the AIDAF framework, the approach can enable users to gain an informed and rapid overview of the Industry 4.0 landscape and characteristics, and to understand and work through the practical steps in alignment with the digital IT strategies in AIDAF [25], which can lead to performing the digital transformation in the above ecosystems. Moreover, given Society 5.0, the author systematized healthcare and manufacturing digital application systems as well as digital platforms in line with digital IT strategies while ensuring information security, privacy, compliance, validation, and other priorities [25], that will lead to performing the digital transformation in the healthcare communities.

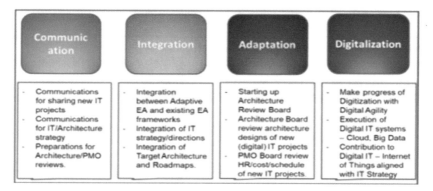

Fig. 1. Digital Transformation Process in the AIDAF Framework – FSAO Approach [24]

4   Cases of Digital Transformation in Enterprise and Ecosystem

4.1   GHE Case in Enterprise to Healthcare Ecosystem (Big Data)

In global EA deployment and digital transformation based on the FSAO approach in GHE, Cloud/Big Data IT strategic projects and systems prioritized in Europe and US group companies were handled by structuring and implementing EA with AIDAF [24]. In this section, characteristics of the digital transformation case with FSAO approach in a global healthcare company is described extended to ecosystem level together with challenges. As enterprise level, the author developed a digital platform for Global Architecture Board (AB) by the AIDAF [24] in Fig. 2, while applying the AIABF to enhance efficiency of AB [24] in Fig. 3. In GHE, the digital transformation was extended with Big Data applications build in data lake using external healthcare professional DB to healthcare ecosystems. We describe the characteristics of the digital transformation in GHE, covering aspects of security and privacy as in the Table 1 below.

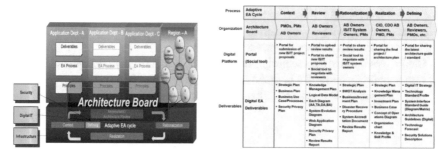

Fig. 2. AIDAF model for GHE [11].   **Fig. 3.** AIABF with GHE [24]

In the GHE, Security Architecture Head joined AB with showing Security guidelines.

Table 1. Characteristics of Digital Transformation in Healthcare Industry

Example Case	Enterprise Level	Ecosystem Level (Station)	Advantages in this Case	Security and Privacy aspects
Global Healthcare Company	- Digital platform for Architecture Board -Digital Transformation with the FSAO approach - Global Architecture Board held by using and applying the AIABF	- Big Data applications with Business Intelligence (BI) covering Healthcare professionals' information	- Efficiency of architecture reviews with digital platform in global AB (Enterprise) - Big Data with external healthcare professional DB, managed in healthcare community. (Ecosystem)	- Security governance was focused with Security Architecture guidelines of cloud and mobile IT applications. (Enterprise) - Security and privacy can be maintained in DA with the data lake related security. (Ecosystem)

4.2 GMC Case (Digital Products) from Enterprise to Ecosystems

In the digital transformation of the GMC, as Communication phase of FSAO approach the global meetings were held to share new IT projects and each IT strategy covering Digital IT for Platform Architecture Board addressing robotic cloud platforms, etc. [32]. In this section, characteristics of the digital transformation case in GMC is described in enterprise toward ecosystem level while accelerating Digital Products of healthcare robots with cloud platforms. As enterprise level, the author started the Digital Platform Board (DPB) with the AIDAF [32] in Fig. 4, while designing the IoRT/IoMRT based robotics digital platform in the AIDAF [32, 33] in Fig. 5. For ecosystem level, digital products of healthcare robots with the digital platforms are accelerated in care centers in healthcare community. We describe the characteristics of the digital transformation in GMC with aspects of security/privacy in Table 2.

Challenges include data security/privacy issues in identifying controller and processor responsibilities and roles in cloud robotics ecosystems [2, 33]. Whereas, Adaptive IoT Security Architecture can contribute to enhance security there [26].

Table 2. Characteristics of Digital Transformation accelerating Digital Products

Example Case	Enterprise Level	Ecosystem Level (Station)	Advantages in this Case	Security and Privacy aspects
Global Manufacturing Company	- Digital Transformation with Digital Platform Board - IoRT/IoMRT based Robotics Digital Platform	- Digital Products of Healthcare robots with Robotics Cloud Platform	- Efficiency in architecture reviews for digital platforms. (Enterprise) - IoMRT-based digital platform can be managed and enhanced in alignment with digital platform strategy. (Enterprise, Ecosystem)	- Challenges exist in identifying controller, processor responsibilities/roles. (Enterprise, Ecosystem) - Adaptive IoT Security Architecture can contribute to enhance security/privacy. (Enterprise, Ecosystem)

Fig. 4. AIDAF model for DPB [32].

Fig. 5. AIDAF model for Medical Robots

4.3   Americas Hospital Case from Enterprise to Ecosystem (Digital Platform)

As the digital transformation in the Americas hospital, they started Communication phase of FSAO approach to share the latest digital IT projects and strategies and designed the prototype of Digital Healthcare Platform (DHP) by Design Thinking approach and built it there [44]. After that, they undertook the design of the above DH for enterprise level, and they plan to extend the DHP to healthcare ecosystem. In this section, characteristics of the digital transformation case in Americas hospital is described in prototype and enterprise level toward ecosystem while designing and building the DHP to improve health status of lifestyle disease patients. As enterprise level, we apply the AIDAF for Design Thinking Approach to ecosystem [44] in Fig. 6, and we show the AIDAF with DHP in Fig. 7 for the ecosystem, as in the Table 3 below.

Moreover, IoT Security functions and 4I framework can contribute to enhance security and privacy there [29], while Digital Trust Framework can be effective [30], too.

Table 3. Characteristics of Digital Transformation in Hospital to Ecosystem

Example Case	Enterprise Level	Ecosystem Level (Station)	Advantages in this Case	Security and Privacy aspects
Americas National Hospital	- Digital Transformation with FSAO/AIDAF for Design Thinking Approach (Prototyping, Enterprise)	- Big Data with Business Intelligence (BI) improving Lifestyle disease and Healthcare status	- Efficiency and Effectiveness in Prototyping and architecture reviews (Enterprise) - Big Data with healthcare information, EHR can contribute to patients care in alignment with Digital IT strategy leading to Society 5.0. (Ecosystem)	- IoT security functions/4I Framework can contribute to enhancing security and privacy. (Enterprise, Ecosystem) - Digital Trust Framework, others will be effective. (Ecosystem)

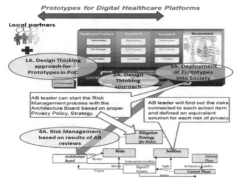

Fig. 6. AIDAF for Design Thinking approach [44]

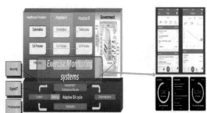

Fig. 7. AIDAF model with DHP [44].

4.4 Smart City Case Directly to Ecosystem

As the digital transformation in Smart City Case in Tokyo, organizations surrounding Tokyo prefecture undertake the projects for prototype of Smart Energy Management system with Digital Platforms with the ecosystem stakeholders such as shopping centers, apartment buildings and city governmental office as shown in Fig. 8. In this section, characteristics of the digital transformation in Smart City Case in Tokyo is described for

Society 5.0 and Industry 4.0 with challenges in Table 4 below. We show the AIDAF with Smart Energy Platform for Design Thinking Approach in Fig. 8 [45] for the ecosystem level, and describe the aspects of security/privacy in Table 4, too.

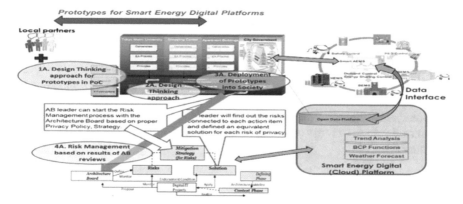

Fig. 8. AIDAF model with Smart Energy Platform for Design Thinking Approach [45]

Table 4. Characteristics of Digital Transformation in Smart City Case

Example Case	Enterprise Level	Ecosystem Level (Station)	Advantages in this Case	Security and Privacy aspects
Smart City in Asia (Energy Industry)	- N/A	- Digital Transformation with AIDAF for Design Thinking Approach (Prototyping, Ecosystem) - AIDAF with Smart Energy Platforms (Ecosystem)	- Efficiency and Effectiveness in Prototyping and architecture reviews (Ecosystem) - Smart Energy digital platforms can be managed in alignment with digital platform strategy of Smart City. (Ecosystem)	- Challenges exist in Global Data Privacy Regulation (GDPR) - Adaptive IoT Security Architecture can contribute to enhance security/privacy. (Ecosystem) - Digital Trust Framework and others will be effective, too. (Ecosystem)

Furthermore, the Adaptive IoT Security Architecture can contribute to enhance security and privacy there [26], while Digital Trust Framework and others will be effective [29, 30], too.

5 Discussion and Challenges

The planning and implementation for digital transformation has many challenges and issues due to the complex nature of global operations in global companies and the many types of digital enterprise strategies and platforms and ecosystems toward Society 5.0 and Industry 4.0. Enterprise architectures being used for supporting such global operations need detailed analysis and appropriate technology solutions implementation. It will be more pragmatic to define the digital transformation process in enterprise toward ecosystem level with FSAO approach in consideration of security and privacy aspects, etc. We propose a more analytical approach to implementation of a digital transformation strategy.

Furthermore, digital IT industry and digitalized society are changing very rapidly. Several case studies toward ecosystem level need to be undertaken and verified in the near future.

6 Conclusion and Next Research

In this paper, we described four cases of performing digital transformation with several examples in enterprises and ecosystems. This research covers both innovative aspects of digital transformation and Security and Privacy perspectives with Risk Management approach there.

Furthermore, we would like to systematize the digital transformation approach in enterprise toward ecosystems with digital platforms in digital strategies in variety kinds of industry with accelerating innovations, while ensuring information security, privacy, etc. from now on.

References

1. Wang, Y., Towara, T., Anderl, R.: Topological approach for mapping technologies in reference architectural model Industrie 4.0 (RAMI 4.0). In: Proceedings of the World Congress on Engineering and Computer Science (2017)
2. Fosch-Villaronga, E., Felzmann, H., Ramos-Montero, M., Mahler, T.: Cloud services for robotic nurses?. In: IEEE/RSJ International Conference on Intelligent Robots and Systems (IROS). Spain. (October, 2018)
3. Guntur, S.R., Gorrepati, R.R., Dirisala, V.R.: Machine Learning in Bio-Signal Analysis and Diagnostic Imaging. Elsevier, 293–318 (2019)
4. Kagermann, H., Wahlster, W., Helbig, J.: Recommendations for implementing the strategic initiative Industrie 4.0. Securing the future of German manufacturing industry [Online] (2013, April)
5. Ross, J.W., Beath, C.M., Mocker, M.: Designed for Digital. The MIT Press, How to Architect Your Business for Sustained Success (2019)
6. Bob, F.: Microservices, IoT and Azure: Leveraging DevOps and Microservice Architecture to Deliver SaaS Solutions. Berkeley, CA: Apr (2015)
7. Boardman, S., Harrington, E.: Snapshot-Open Platform 3.0™. The Open Group (2015)
8. Alwadain, A., Fielt, E., Korthaus, A., Rosemann, M.: A comparative analysis of the integration of SOA elements in widely-used enterprise architecture frameworks. Int. J. Intell. Inf. Technol. **9**(2), 54–70 (2014)

9. Buckl, S., Matthes, F., Schulz, C., Schweda, C.M.: Exemplifying a framework for interrelating enterprise architecture concerns. In: Sicilia, M.A., Kop, C., Sartori, F. (eds.) ONTOSE 2010. LNBIP, vol. 62, pp. 33–46. Springer, Heidelberg (2010). https://doi.org/10.1007/978-3-642-16496-5_3

10. Masuda, Y., Shirasaka, S., Yamamoto, S., Hardjono, T.: International journal of enterprise information systems JEIS. IGI Global **13**, 1–22 (2017)

11. Masuda, Y., Shirasaka, S., Yamamoto, S., Hardjono, T.: Architecture board practices in adaptive enterprise architecture with digital platform: a case of global healthcare enterprise international journal of enterprise information systems. IGI Global **14**, 1 (2018)

12. Aceto, G., Persico, V., Pescapéa, A.: The role of information and communication technologies in healthcare: taxonomies, perspectives, and challenges. J. Netw. Comput. Appl. **107**, 125–154 (2018)

13. Asif Qumer G. Adaptive cloud Enterprise Architecture. Intelligent Information Systems 4 Singapore: World Scientific Publishing Co (2015)

14. Archenaa, J., Anita, E.M.: A survey of big data analytics in healthcare and government. Procedia Comput Sci **50**, 408–413 (2015)

15. Chawla, N.V., Davis, D.A.: Bringing big data to personalized healthcare: a patient-centered framework. J. Gen. Intern. Med **28**, 660–665 (2013)

16. Osmani, V., Balasubramaniam, S., Botvich, D.: Human activity recognition in pervasive health-care: supporting efficient remote collaboration. J Netw Comput Appl **31**, 628–655 (2008)

17. Khan, K.M., Gangavarapu, N.M.: Addressing cloud computing in enterprise architecture: issues and challenges. Cutter IT J. **22**(11), 27–33 (2009)

18. Patel, P., Cassou, D.: Enabling High-level Application Development for the Internet of Things. J. Syst. Softw. 1–26 (2015). Elsevier

19. Iacob, M.E., et al.: Delivering Business Outcome with TOGAF® and ArchiMate®: BiZZdesign (2015)

20. Johnson, P., et al.: IT Management with Enterprise Architecture Stockholm: KTH (2014)

21. The Open Group. TOGAF Version 9.1: Van Haren Publishing (2011)

22. Zimmermann, A., Schmidt, R., Sandkuhl, K., Jugel, D.: Digital enterprise architecture – transformation for the Internet of Things. In: Enterprise Distributed Object Computing Workshop (EDOCW), IEEE 19th International (2015)

23. Couturier, J., Sola, D., Borioli, G.S., Raiciu, C.: How can the Internet of Things help to overcome current healthcare challenges. Commun. Strat **87**, 67–81 (2012)

24. Masuda, Y., Zimmermann, A., Bass, M.: Adaptive enterprise architecture process for global companies in a digital IT era. Int. J. EIS **17**(2), 21–43 (2021). IGI Global

25. Masuda, Y., Zimmermann, A., Sandkuhl.: Applying AIDAF for enabling industry 4.0 in open healthcare platform 2030. In: Proceedings of the 14th International KES Conference on Human Centred Intelligent Systems (2021)

26. Gill, A.Q., Beydoun, G., Niazi, M., Khan, H.U.: Adaptive architecture and principles for securing the IoT systems. In: Barolli, L., Poniszewska-Maranda, A., Park, H. (eds.) IMIS 2020. AISC, vol. 1195, pp. 173–182. Springer, Cham (2020). https://doi.org/10.1007/978-3-030-50399-4_17

27. Nayyar, A., Batth, R.S., Nagpal, A.: Internet of Robotic Things: driving intelligent robotics of future- concept, architecture, applications and technologies. In: 4th IEEE International Conference, pp. 151–160 (2018)

28. BITKOM, VDMA, and ZVEI (2015, April). Umsetzungsstrategie Industrie 4.0 – Ergebnisbericht der Plattform Industrie 4.0. https://www.bitkom.org/Publikationen/2015/Leitfaden/Umsetzungsstrategie-Industrie-40/150410-Umsetzungsstrategie-0.pdf

29. Dasgupta, A., Gill, A., Hussain, F.: A conceptual framework for data governance in IoT-enabled digital IS ecosystems. In: Proceedings of the 8th International Conference on Data Science, Technology and Applications (DATA 2019), pp. 209–216 (2019)
30. Information Systems Audit and Control Association (ISACA).: New ISACA Guide Outlines Key Components of Digital Trust Implementation (May 2022). https://www.isaca.org/why-isaca/about-us/newsroom/press-releases/2022/new-isaca-guide-outlines-key-components-of-digital-trust-implementation
31. Masuda, Y., Shepard, D.S., Yamamoto, S., Toma, T.: Clinical decision-support system with electronic health record: digitization of research in pharma. In: Chen, Y.-W., Zimmermann, A., Howlett, R.J., Jain, L.C. (eds.) Innovation in Medicine and Healthcare Systems, and Multimedia. SIST, vol. 145, pp. 47–57. Springer, Singapore (2019). https://doi.org/10.1007/978-981-13-8566-7_5
32. Masuda, Y., Zimmermann, A., Shirasaka, S., Nakamura, O.: Internet of robotic things with digital platforms: digitization of robotics enterprise. In: Zimmermann, A., Howlett, R.J., Jain, L.C. (eds.) Human Centred Intelligent Systems. SIST, vol. 189, pp. 381–391. Springer, Singapore (2021). https://doi.org/10.1007/978-981-15-5784-2_31
33. Masuda, Y., Shepard, D.S., Nakamura, O., Toma, T.: Vision paper for enabling internet of medical robotics things in open healthcare platform 2030. In: Chen, Y.W., Tanaka, S., Howlett, R.J., Jain, L.C. (eds.) Innovation in Medicine and Healthcare. SIST, vol. 192, pp. 3–14. Springer, Singapore (2020). https://doi.org/10.1007/978-981-15-5852-8_1
34. Garnier, J.-L., Bérubé, J., Hilliard, R.: Architecture Guidance Study Report 140430. ISO/IEC JTC 1/SC 7 Software and systems engineering. (2014)
35. Tamm, T., Seddon, P.B., Shanks, G., Reynolds, P.: How does enterprise architecture add value to organizations? Commun. Assoc. Inf. Syst. **28**, 10 (2011)
36. Chen, H.-M., Kazman, R., Perry, O.: From software architecture analysis to service engineering: an empirical study of methodology development for enterprise SOA implementation. IEEE Trans Serv Comput **3**, 145–160 (2014)
37. MacKenzie, C.M., Laskey, K., McCabe, F., Brown, P.F., Metz, R.: Reference Model for SOA 1.0. (Technical Report). Advancing Open Standards for the Information Society (2006)
38. Newman, S.: Building Microservices. O'Reilly Media, Sebastopol (2015)
39. Richards, M.: Microservices vs. Service-Oriented Architecture.1st Edn. O'Reilly Media, Sebastopol (2015)
40. Muhammad, K., Khan, M.N.A.: Augmenting mobile cloud computing through enterprise architecture: survey paper. Int. J. Grid Distrib. Comput. **8**, 323–336 (2015)
41. Gill, A.Q., Smith, S., Beydoun, G., Sugumaran, V.: Agile enterprise architecture: a case of a cloud technology-enabled government enterprise transformation. In Proceedings of the 19th Pacific Asia Conference on Information Systems (PACIS), pp. 1–11 (2014)
42. Masuda, Y., Shirasaka, S., Yamamoto, S.: Integrating mobile IT/Cloud into enterprise architecture: a comparative analysis. In: Proceedings of the 21th Pacific Asia Conference on Information Systems (PACIS), p. 4 (2016)
43. Deguchi, A., Hirai, C., Matsuoka, H., Nakano, T.: Society 5.0. Springer, Singapore (2020) https://doi.org/10.1007/978-981-15-2989-4
44. Masuda, Y., Zimmermann, A., Schmidt, R.: An adaptive enterprise architecture design for a digital healthcare platform: toward digitized society–Industry 4.0, Society 5.0. In: Enterprise Distributed Object Computing Workshop (EDOCW), IEEE 25th International (2021)
45. Masuda, Y., Jain, R.: Vision paper for sharing economy and digital platforms toward society 5.0. In: International KES Conference on Human Centered Intelligent Systems (2022)

Author Index

© The Editor(s) (if applicable) and The Author(s), under exclusive license
to Springer Nature Singapore Pte Ltd. 2023
A. Zimmermann et al. (Eds.): KES-HCIS 2023, SIST 359, pp. 177–178, 2023.
https://doi.org/10.1007/978-981-99-3424-9

Printed in the United States
by Baker & Taylor Publisher Services